MW01633963

"*Intelligence Communication in the Digital Era* is essential to every strategy, marketing, finance, and intelligence professional who understands that professional career viability is dependent on the ability to navigate streams of information and associated decision delivery systems. The book tells how to convey critical recommendations and executive support in clear, concise, and creative ways; capture attention at the highest levels; and provides easy access to decision support criteria that give organizations a competitive advantage. Kudos to Arcos and Pherson for authoring a resource that provides application rather than theory to the overarching requirements placed on intelligence professionals in a world where information abounds." – Nanette Bulger, Executive Director and CEO, Strategic and Competitive Intelligence Professionals (SCIP)

"In an age of complexity, velocity, and high jeopardy, the challenge to exert effect has never been greater for an information exploitation specialist – be it a business analyst, an insurance actuary, or an intelligence officer managing national security requirements ranging from defence to terrorism. Rethinking the fundamentals of assessment methodology, and particularly the conveyance of the message, will be the key to success for those in the analytical sphere. In a world where decision-makers are overwhelmed by data, increasingly secure in their own belief systems, and cynicism or distrust grow stronger with each alleged scandal, the analyst must become adept at understanding client needs and able to offer clear, well-founded, and effectively marketed judgments. This is not about distorting or perverting the sanctity of the objective message; rather, it is acknowledging the impact of a fast-paced business environment as well as recognizing what is emerging as an increasingly prevalent and distortive cognitive bias that exists within all of us due to this Internet-enabled information age." – Ray Boisvert, Former Assistant Director, Intelligence, Canadian Security Intelligence Service and President/CEO, I-Sec Integrated Strategies

"*Intelligence Communication in the Digital Era* reminds us of two very important things: technological change is about delivering intelligence as well as gathering it; and change is as much an opportunity as a risk. Arcos and Pherson continue to lead the thinking in this area for academics and practitioners alike with this highly significant new contribution to the Intelligence Studies literature." – Julian Richards, Co-Director, Centre for Security and Intelligence Studies (BUCSIS), University of Buckingham, UK

"Although the intelligence community is an information industry, in many ways its processes have changed little since the digital revolution. To avoid obsolescence, intelligence organizations will need to modernize by reforming information management, promoting asynchronous collaboration, and adopting a model of production not based on the 'document' paradigm. This book can guide that effort; it will be a valuable resource for intelligence practitioners and their managers in both industry and government." – Nick Hare, Former head of Futures and Analytical Methods, UK Defence Intelligence, UK

"Technological advances are dramatically impacting what information is available to the community of policy and decision makers, where they get it from, when they want to view it, and how they make sense of it. These changes exacerbate half-century-old disconnects between intelligence providers and those they would serve, making it harder to overcome barriers imposed by outmoded technology and inadequate techniques for facilitating well-advised decisions. This book opens a valuable window on this world of challenges, starting with an astute introduction to the landscape by Pherson and Arcos, followed by six thoughtful chapters on significant issues in communications, collaboration, and producer/consumer relations. It is a compendium of excellent insights with lots of supporting references pointing to a new approach." – Robert Neches, Former Director of Incisive Analysis, Intelligence Advanced Research Projects Activity (IARPA), Office of the Director of National Intelligence, USA

DOI: 10.1057/9781137523792.0001

Other Publications

Rubén Arcos

William J. Lahneman and Rubén Arcos, (eds), *The Art of Intelligence: Simulations, Exercises, and Games*. (Lanham, Maryland: Rowman & Littlefield Publishers, 2014) (Security and Professional Intelligence Education Series).

Fernando Velasco and Rubén Arcos, (eds), *Estudios en Inteligencia: respuestas para la gobernanza democrática*. (Madrid: Plaza y Valdés/Ministerio de la Presidencia, 2014).

Arcos, Rubén, *La lógica de la excepción cultural. Entre la geoeconomía y la diversidad cultural.* (Madrid: Cátedra Signo e Imagen, 2010).

Arcos, Rubén, "Systems of Intelligence: Spain" in Robert Dover, Michael Goodman and Claudia Hillebrand, (eds), *Routledge Companion to Intelligence Studies*. (London and New York: Routledge, 2013) pp. 235–242.

Arcos, Rubén, "Academics as Strategic Stakeholders of Intelligence Organizations: A View from Spain". *International Journal of Intelligence and Counterintelligence*. 26 (2), (2013), pp. 332–346.

Arcos, Rubén, "Intelligent Design – Restructuring the Spanish Security Apparatus". *Jane's Intelligence Review*. 24 (8) (August 2012), pp. 36–39. First published online at jir.janes.com on 29 June 2012.

Randolph H. Pherson

Randolph H. Pherson, *Handbook of Analytic Tools and Techniques, 4th Edition*. (Reston, VA: Pherson Associates, LLC, 2015).

Richards J. Heuer Jr. and Randolph H. Pherson, *Structured Analytic Techniques for Intelligence Analysis, 2nd Edition*. (Washington, DC: CQ Press/Sage Publications, 2015).

Sarah Miller Beebe and Randolph H. Pherson, *Cases in Intelligence Analysis: Structured Analytic Techniques in Action, 2nd Edition*. (Washington, DC: CQ Press/Sage Publications, 2015).

Randolph H. Pherson and Louis M. Kaiser, *Analytic Writing Guide*. (Reston, VA: Pherson Associates, LLC, 2014).

Katherine H. Pherson and Randolph H. Pherson, *Critical Thinking for Strategic Intelligence*. (Washington, DC: CQ Press/Sage Publications, 2013).

Randolph H. Pherson, "The New Age of Structured Analytic Techniques" in George, Roger Z. and James B. Bruce, (eds), *Analyzing Intelligence: Origins, Obstacles, and Innovations, 2nd Edition*. (Washington, DC: Georgetown University Press, 2014).

Randolph H. Pherson, "Five Habits of the Master Thinker". *Journal of Strategic Security*, 6 (3), (Fall 2013).

palgrave▸pivot

Intelligence Communication in the Digital Era: Transforming Security, Defence and Business

Edited by

Rubén Arcos

Professor, Center for Intelligence Services and Democratic Systems, Rey Juan Carlos University, Madrid, Spain

and

Randolph H. Pherson

President, Pherson Associates, Reston, Virginia, United States

DOI: 10.1057/9781137523792.0001

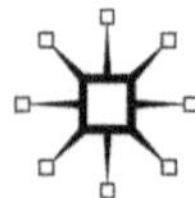

First published 2015
PALGRAVE MACMILLAN

Palgrave Macmillan in the UK is an imprint of Macmillan Publishers Limited, registered in England, company number 785998, of Houndmills, Basingstoke, Hampshire RG21 6XS.

Palgrave Macmillan in the US is a division of St Martin's Press LLC, 175 Fifth Avenue, New York, NY 10010.

Palgrave Macmillan is the global academic imprint of the above companies and has companies and representatives throughout the world.

Palgrave® and Macmillan® are registered trademarks in the United States, the United Kingdom, Europe and other countries.

ISBN: 978-1-137-52380-8 EPUB
ISBN: 978-1-137-52379-2 PDF
ISBN: 978-1-137-52378-5 Hardback

A catalogue record for this book is available from the British Library.

A catalog record for this book is available from the Library of Congress.

www.palgrave.com/pivot

DOI: 10.1057/9781137523792

To those who see the need for flexibility and agility and approach the future with an open mindset, seeking out opportunities where others see only challenges.

All statements of fact, opinion, or analysis expressed in the Introduction, Chapter 2, and Chapter 4 are those of the respective authors and do not reflect the official positions of the Central Intelligence Agency or any other US government agency. Nothing in the contents should be construed as asserting or implying US government authentication of information or agency endorsement of the author's views. These sections have been reviewed by the Central Intelligence Agency only to prevent the disclosure of classified information.

DOI: 10.1057/9781137523792.0001

Contents

DOI: 10.1057/9781137523792.0001

List of Figures

DOI: 10.1057/9781137523792.0002

DOI: 10.1057/9781137523792.0002

Foreword

This book should be required reading for almost anyone in the intelligence field. As the world becomes increasingly complex, technology constantly redefines how we think, and the 24/7 news cycle dictates our patterns of behavior, the intelligence community can no longer afford to deliver its analysis only in printed format based on a 24-hour – or more often multiple-day – production cycle. We need to embrace new technologies both in crafting our message and delivering it to the policymaker. *Intelligence Communication in the Digital Era* lays out the framework for understanding how best to accomplish this mission to transform the production of intelligence analysis from a static, narrative product to a more dynamic, digitalized, and potentially interactive format.

I have long argued that the core responsibility of an intelligence analyst is to provide to the policymaker the most accurate and objective information possible. Our concept of what constitutes good policy support, however, is undergoing a major transformation. Delivering a tightly organized, crisply written narrative is no longer the primary requirement. Our customers are living in a much more complex world where visual images often convey needed information far more effectively than text. This presents a challenge to the analyst in terms of both collection and presentation.

Sources always need to be verified and this can pose real problem in processing images. For example, pictures taken by cell phones now can easily be digitally altered. Photos are more frequently being purposely mislabeled, for example, by misrepresenting a video of the effects of an airstrike as having happened yesterday when it actually occurred six months ago.

The second challenge poses an even more vexing problem. Pictures and videos not only convey information but can also stir emotions. How does an analyst decide if a picture "lacks objectivity" because it induces too strong an emotional reaction? The intelligence community has well-developed standards for determining what is appropriate to say and not say in a written document but has barely begun to address this question when it comes to the use of videos, simulations, and other visual interfaces.

Intelligence Communications in the Digital Era approaches these issues from an international perspective. Authors from the United States, Canada, and Spain offer valuable insights from a broad range of perspectives ranging from communicating risk to enhancing collaboration to leveraging the use of modeling and simulation. The authors also come from a variety of backgrounds in academia, intelligence, and the business world. This is a ground-breaking book that has much to say not only to intelligence professionals but to analysts and managers of analysts in any domain.

Charles E. Allen

Former Under Secretary for Intelligence and Analysis,
Department of Homeland Security

Former Assistant Director of Central Intelligence for Collection,
Central Intelligence Agency

DOI: 10.1057/9781137523792.0003

Notes on Contributors

Rubén Arcos is Professor of Communication Sciences at Rey Juan Carlos University in Madrid, Spain. He also serves as deputy director, Centre for Intelligence Services and Democratic Systems and coordinator of the Master Program in Intelligence Analysis. Dr Arcos is founder and chair of the Strategic and Competitive Intelligence Professionals (SCIP) Chapter in Spain and deputy editor of the Spanish academic intelligence journal *Inteligencia y seguridad: Revista de análisis y prospectiva.* His last book is *The Art of Intelligence: Simulations, Exercises and Games.* His main research interests are strategic communications, intelligence analysis, and experiential learning.

Jonathan Calof is Professor of Strategy and International Business and executive editor for *Frontline Safety and Security Magazine*. He combines research and consulting in competitive intelligence, technical foresight, and business analytics to help organizations develop insights on their competitive environment. He has over 150 publications and has given over 1,000 speeches, seminars, and keynote addresses on intelligence and foresight and has helped several companies and government agencies enhance their intelligence capabilities. A recipient of Frost and Sullivan's life time achievement award in competitive intelligence, he was also named a fellow of the Society of Competitive Intelligence Professionals (SCIP).

Aaron B. Frank is an information scientist at RAND. His scholarly and professional work has focused on the use of social science theories, gaming, simulation and computation in strategic assessments, intelligence analysis, and managing risk and uncertainty. He currently supports research projects on strategic foresight, integrating computational modeling and simulation into analytic tradecraft, and developing and assessing future concepts of operations for several US government sponsors. He holds a BA in Political Science from Boston University, an MA in National Security Studies from Georgetown University, and a PhD in Computational Social Science from George Mason University.

Mary O'Sullivan is Dean of the Forum at Pherson Associates. She designs curricula and specialized workshops for analysts in the private and public sectors. She has developed classes in critical thinking, structured analytic techniques, and analytic writing and taught students across the US Intelligence Community. O'Sullivan has also co-developed several analytic techniques useful in conducting foresight analysis. A career CIA intelligence officer, she was the first Chancellor of CIA University. She last served as Deputy Director of the Office of Policy Support. She received her BA from Western Kentucky University and her MA in History from the University of Kansas.

Randolph H. Pherson is President, Pherson Associates, LLC and CEO, Globalytica, LLC. He teaches analytic techniques and critical thinking skills in the United States and abroad. Pherson co-authored *Structured Analytic Techniques for Intelligence Analysis, Cases in Intelligence Analysis: Structured Analytic Techniques in Action, Critical Thinking for Strategic Intelligence, and Analytic Writing Guide.* He wrote the *Handbook of Analytic Tools and Techniques* and holds the patent on a suite of analytic tools, TH!NK Suite®. A career CIA intelligence analyst, he last served as National Intelligence Officer for Latin America. He received his AB from Dartmouth College and his MA in International Relations from Yale University.

John Pyrik has a broad range of analytical and investigative experience accumulated from 25 years of government service. In 2005, he was visiting fellow, Canadian Centre of Intelligence and Security Studies, Carleton University and became chief instructor/coordinator for community-wide training of Canadian intelligence analysts. In 2011, he won an award from the Canadian Association of Professional Intelligence Analysts for "Advancing the Tradecraft of Analysis." He has served on the boards of the Canadian Association for Security and Intelligence Studies, International Association for Law Enforcement Intelligence Analysts, and International Association for Intelligence Education and presently instructs at the Justice Institute of British Columbia.

DOI: 10.1057/9781137523792.0004

Introduction: The Changing Intelligence Communications Landscape

Randolph H. Pherson and Rubén Arcos

Arcos, Rubén and Randolph H. Pherson, eds. *Intelligence Communication in the Digital Era: Transforming Security, Defence and Business.* Basingstoke: Palgrave Macmillan, 2015. DOI: 10.1057/9781137523792.0005.

Emerging information and communication technologies promise to fundamentally change how analysis is produced and used by key customers in the coming years. A major transition is just getting underway as producers of analytic products shift from delivering static, hardcopy narrative papers to relying increasingly on more dynamic, digitally-based modes of presentation. Tomorrow's intelligence consumer will be increasingly inclined to seek information and analytic insights in digital and interactive formats. The ability of producers of analysis to adapt intelligence deliverables meet this challenge could well determine whether such analytic units remain competitive in an era of digital communication.

This volume is a collection of seven papers, drafted by authors from the United States, Canada, Spain, and the UK.[1] In the book, we will explore how to:

- Present analysis to – and customize information for – senior decision makers and policymakers.
- Communicate information differently because of advances in technology.
- Encourage analysts to work differently and provide them with the resources and production technology to do so.
- Invest in new tradecraft that prioritizes developing producer-consumer relations rather than predicting the future or revealing the "truth" devoid of policymaker's context or interests.

Specifically, the book is intended to:

- Focus on an emerging theme that as technology improves the delivery of information and analysis will be moving from push to pull methods.
- Help analysts and managers of analysts rethink how to access as well as present data, analysis, and key findings.
- Suggest new ways to facilitate quick access to data, analysis, and key findings.
- Provide examples of how to extend communication beyond narrative to audio, video, and other social media formats.
- Describe ways to communicate with more impact.

New technologies

The coming transition from static to more dynamic modes of delivery of information and analysis is spurred by the public's growing familiarity

DOI: 10.1057/9781137523792.0005

with hypermedia, hypertext, interactive multimedia contents, immersive communications, responsive websites, and data visualizations tools. The widespread use of portable devices and multi-touch screens reinforces this trend. Similarly, the use of interactive visualization tools and big data analytics in business marketing to organize and display intelligence about customers is a growing trend. Interactive multimedia and computerized simulations offer a dazzling array of options for communicating analytic judgments to intelligence customers in an effective manner. In sum, technology assists immediacy, portability, and offers freedom from location constraints; it can also foster collaboration and make customization a realizable, cost-efficient goal.

The emergence of exciting, new technologies such as augmented reality and wearable devices challenge the ways in which information and analysis can be presented to decision makers. Delivery mechanisms that use telepresence, augmented reality, and computerized glasses offer major opportunities to redesign the producer-consumer interface and have only just begun to be explored. One caveat: the communicative opportunities derived from these technologies can also create serious information security risks and present unanticipated challenges in terms of adaptation and the development of knowledge and skills.

New generation of users

The new generation of users will have a totally different set of expectations than the traditional consumers of analysis today. They will want to be in control of how they become informed; and to own it. We believe they will increasingly demand the freedom to:

- Seek information and analysis at any time; they do not want to be straitjacketed because they can only receive a briefing or read a daily report at a set time each day and only once each day.
- Decide where to focus their attention and not be a slave to a table of contents or an order of presentation.
- Receive updates on key topics that they can access or ignore depending on their schedule.

In essence, this will require a major shift from traditional "push" systems of delivering analytic products to "pull" systems that allow the customer to navigate a wide range of messages and delivery options, focusing only

DOI: 10.1057/9781137523792.0005

on what interests them the most. No longer will they be dependent on the standard morning briefing or the delivery of pre-packaged analysis contained in a static, hard-copy report. Ideally, they might even have the opportunity to ask questions or craft quick turn-around queries as they work their way around the digital display.

One of the benefits of a "pull" system is that much larger stores of information can be captured on a tablet, but displayed only if specifically requested. Links can be embedded in the text that – when clicked on by a mouse – can open up more detailed files or reports that provide background information or display what was previously written (or, in this new world, should we say "prepared"?) on that topic. This gives the customer the ability to drill down in a particular area of key interest but also the freedom to ignore that information entirely should the customer choose to focus attention on other topics.

Another benefit is that more dynamic systems of information exchange facilitate the ability to match queries and responses to a customer's specific needs. With the use of pull-down menus, for example, customers could identify their priority areas of concern and only process information and analysis relevant to those topics. A downside risk to this approach, however, is that the customer becomes too narrowly focused, ignoring other major developments that might have secondary or tertiary impact on their world.

New time-management possibilities

The movement from a static, "push" system based on scheduled briefings or hard copy reports to a more dynamic "pull" system that can be accessed on demand and continuously updated has major implications for time management. With a "pull" system, the customer is no longer locked into a set period of time to be briefed or to read each day. He or she can *vary* how much time is devoted to the acquisition of information and analysis as well as how much time should be devoted to reflect on those inputs depending on the time and scheduling demands of that particular day.

The customer can also dictate *when* he or she processes the information. Some customers prefer to start each day with a morning update or morning brief while others approach the task of becoming informed more strategically, allocating time later in the day to get their inputs.

A key challenge in developing a dynamic, on-demand system of information delivery is not to overwhelm the customer with data. In a

DOI: 10.1057/9781137523792.0005

briefing environment, the customer usually can comment on a particular report, reflect on its significance, and engage the briefer in a dialogue about the topic. An on-demand system, however, is not likely to possess such a dialogue feature. A chat room might be offered, but most customers would view this as a cumbersome and overly time consuming way to engage in a conversation. More satisfying arrangements such as one-on-one video (Skype) or avatar-based collaboration platforms (TH!NK Live™) could be used to support such synchronous conversations, but scheduling issues could pose some obstacles.[2]

New media

The delivery of information and analysis on a dynamic, digitally-based platform offers the benefit of introducing new types of media to the customer.

- **Video clips** can be included to help prepare a customer for an upcoming meeting with another senior official. The video could show the policymaker interacting with another or reveal how a particular affliction has affected their gait or how a recent stroke has affected their speech. Videos captured on YouTube or other social media sites can quickly convey the severity of a bombing or suicide attack or the intensity of emotions exhibited by demonstrators.
- **Infographics** can provide easily digestible, visual representations of complex information, data, or knowledge quickly and clearly. By utilizing graphics, they enhance the brain's ability to see patterns and trends (see Figure 1.1).
- **Maps** can be displayed, giving the user the opportunity to zoom in on a particular structure or geographic feature or to zoom out to capture the broader landscape. Maps can also be annotated to explain key terrain features or highlight key facilities, roads, or other lines of communication.
- **Overhead imagery** can depict the size of a popular demonstration or the presence of a weapons system. Pictures of the imaged weapons system could be pasted on top of the imagery. Some customers might even be given the opportunity to conduct video "flyovers" of the landscape, viewing the terrain as if they were in an airplane or helicopter.
- **Animations** can be created to let the data "speak for itself." For example, a dynamic map could be created showing how a

DOI: 10.1057/9781137523792.0005

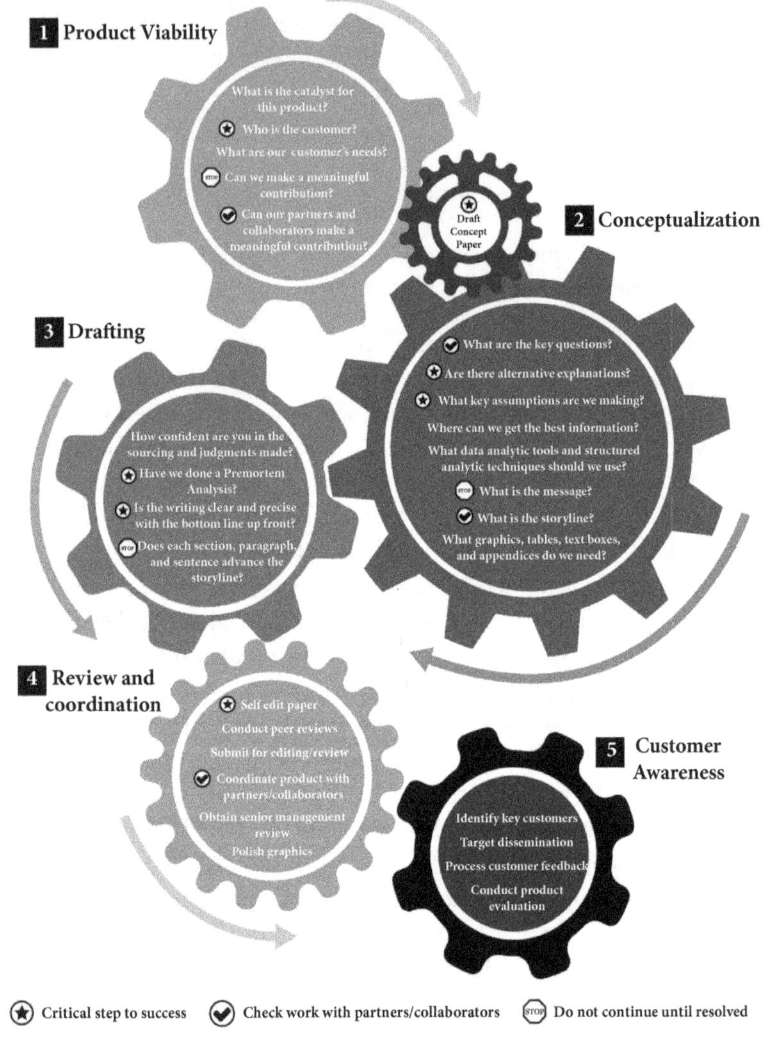

FIGURE I.1 *Multi-organization analytic production process*

Source: Globalytica, LLC, 2015

DOI: 10.1057/9781137523792.0005

contagious disease is starting to spread through a population or how a growing number of troops are being deployed to various points along a border.
- **Graphs, charts, and tables** can be made more digestible by allowing the consumer to hide columns or manipulate summary statistics of the data shown in a matrix.

Customers often ask to see the raw, unevaluated intelligence reports that comprise the sources on which more finished analytic products are based. With a tablet delivery system, such raw reporting can easily be coupled with a quick analysis of:

- How the report should be interpreted.
- The accuracy of the data.
- The quality of the sourcing.
- The analysts' level of confidence in their key judgments or key findings.

Such data could be appended to the report and show up on the tablet screen as a pop-up box or as commentary posted on the bottom or on the side of the screen.

Another potential benefit of moving from static, hard copy reports to a more dynamic, digitally-based presentation of information and analysis is the ability to dramatically increase the amount of available information through the insertion of hyperlinks in the narrative. For example, if a person is mentioned in the text, a click on their name would reveal their biographic summary or a click on a town name could reveal a map showing its location.

Care must be taken, however, to avoid overwhelming the reader with reams of additional information that the reader cannot usefully absorb. Most readers are seriously time constrained, and hyperlinks should not be used to justify loading the customer down with ten times more data than he or she could ever absorb. Similarly, care must be taken not to clutter the screen with sidebars and optional drawdowns diverting attention from the main message of the presentation.

New frameworks for analysis

One of the greatest – and largely unrealized – advantages of moving to a digital-based presentation of information and analysis is the ability to

DOI: 10.1057/9781137523792.0005

provide better frameworks for understanding that data. For example, much of the analysis we read every day is based on sets of assumptions that are never explicitly stated and can often be wrong. With a tablet, it would be fairly simple to provide a standardized icon (in the far right hand corner, for example) labeled Key Assumptions. When the user clicked on the icon, a drop down box would appear providing a list of key assumptions the analyst or team of analysts were making when they produced the analytic product. Most analysts recognize the need to identify and challenge their key assumptions when preparing a draft assessment. In fact, when one of the contributing authors was serving as National Intelligence Officer for Latin America, he was required to include a Key Assumptions text box with every National Intelligence Estimate he published. If this became a standard practice for most analytic products, the overall quality of analytic products would almost certainly increase.

Similar dropdown boxes could be created for a variety of tasks informing the reader about:

Critical information gaps. In researching this paper, did the author have difficulty finding information on any particular topic? How might this have affected the analysis?

The So What? (and the So What of the So What?). For the busy reader, a short box that succinctly describes for the reader why this trend or event matters would be highly welcome. Even more appreciated would be a box that outlines the secondary or tertiary implications of that same trend or event. For example, in an article describing Russian behavior after Malaysian Air 17 was shot down over the Ukraine, a drop down box might say:

Is Russia responsible for the downing of MH17?

Video images of a Russian BAK missile launcher system [without one of its missiles] crossing the border into Russia from part of the Ukraine where the Malaysia Airlines flight MH17 was shot down provides circumstantial evidence that the Russian separatists were responsible for the shoot down (the So What?). The image buttresses arguments that Russian separatists may have been assisted by Russian technicians in launching the missile. Russia may now be less disposed to provide such systems to Russian separatists (the So What of the So What?).

DOI: 10.1057/9781137523792.0005

Indicators. If the assessment describes potential future scenarios, then a useful drop down box would contain a list of indicators the reader could monitor to determine whether such a scenario was beginning to emerge.

Opportunities. The purpose of an analytic product is to explain what has occurred and explore what might evolve in the future. The value of the product can be greatly enhanced if the drafter can go an extra step and suggest how the reader (often a key decision maker or policymaker) could take specific actions to help a good scenario come to pass or prevent a bad scenario from happening. These suggestions could be inserted into a drop down box that the reader could easily access if he or she was interested in such advice.

This book does not portend to propose a unique and standardized model of delivering multimedia products or to advocate for the end of hard copy narrative products. Rather, it aspires to explore the challenges inherent in looking at intelligence analysis through a new set of "digital communication" glasses and to suggest new methods for presenting information and analysis in ways that better respond to the needs and expectations of tomorrow's intelligence consumer.

Notes

1 The inspiration to write this book came after many of the authors presented papers on this topic for a panel entitled Reinventing Intelligence Production for the 21st Century at the International Studies Association's 55th Annual Convention in Toronto, Canada in March 2014.

2 For more information on the TH!NK Live™ avatar-based virtual learning environment go to www.globalytica.com.

DOI: 10.1057/9781137523792.0005

1

Communicating Analysis in a Digital Era[1]

Rubén Arcos

Abstract: *Communication is one of the cornerstones of the intelligence process, and recent developments in multimedia communication are likely to have a major impact on how intelligence analysis is presented to the user of the intelligence product. Digital communication is driving deep changes in the way information is produced and consumed globally and across industries, posing challenges to how intelligence analysis is delivered from the perspective of both the producer and the consumer. These changes will affect the concepts of usability, user experience, interaction design, and information design. Although intelligence agencies have their own distinctive features in relation to intelligence clients at the corporate level as well as certain security standards and counterintelligence imperatives to guarantee, they will need to adapt their intelligence products to the digital era to remain competitive.*

Keywords: intelligence analysis; intelligence product; multimedia communication; user experience

Arcos, Rubén and Randolph H. Pherson, eds. *Intelligence Communication in the Digital Era: Transforming Security, Defence and Business.* Basingstoke: Palgrave Macmillan, 2015. DOI: 10.1057/9781137523792.0006.

The changing communications environment

In the intelligence studies literature, academic discussion, and even professional practice, the communication of analytic products to government decision makers as a step in the intelligence process has traditionally received much less attention than intelligence collection and analysis. Without the effective communication of intelligence to policymakers, however, all the previous efforts in collection and analytic production are futile. This process involves activities that require the acquisition of specific competencies to be conducted successfully.

The communication domain is driving huge transformations around the world in virtually all industries, changing the ways we interact, build, and conduct relations with others. It is difficult to imagine a field of human experience that has not been transformed by digital information and communication technologies. The editorial staffs of newspapers, news agencies, and other traditional media all over the world have witnessed and are experiencing major transformations driven by technological innovations while they struggle to adapt the processes of producing and delivering news and information of relevance to their readers.

Multimedia communication among individuals is nowadays usual in both professional and private spheres. Multimedia can be defined as "any combination of text, graphics, video, audio, and animation in a distributable format that consumers can interact with using a digital device."[2]

Newspapers provide highly attractive interactive infographics in their digital editions, presenting data and information in a digestible format. Videos increasingly are being introduced in digital publications. People are growing increasingly familiar with digital technologies and are both consumers and producers of digital photography, video, and blogging and micro-blogging platforms. They are consumers of information but are also learning how to produce content for others.

Coincident with this ongoing transformation, people are getting used to new ways of consuming information and interacting with that information. They are assuming a more active role in the process. Consequently, decision makers in government and industry are demanding – and will increasingly demand – analysis and intelligence products adapted to this new era of digital communication. As the threshold of usefulness for new information decreases, the consumer of intelligence analysis is expecting shorter timeframes for receiving intelligence products (see Figure 1.1).

DOI: 10.1057/9781137523792.0006

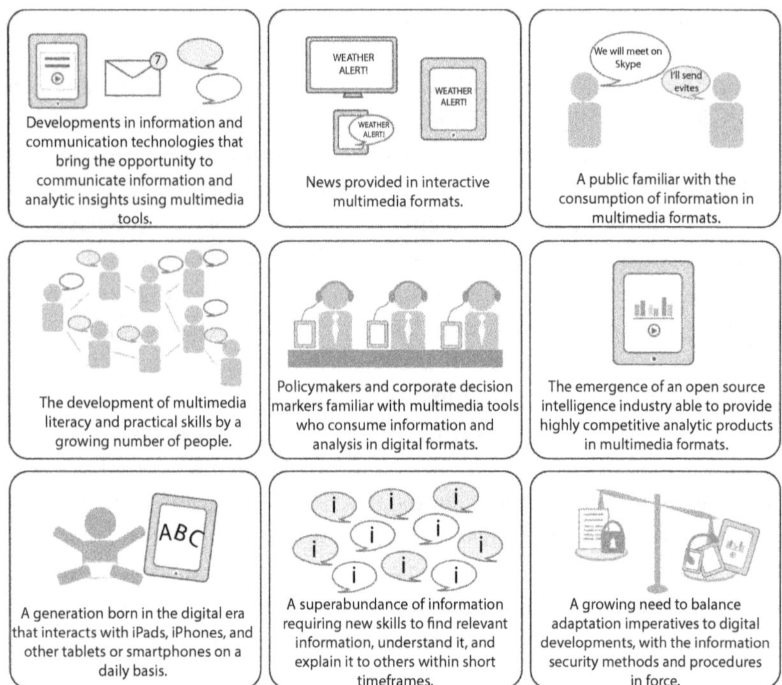

FIGURE 1.1 *Characteristics of the emerging communications environment*

The intelligence user experience (UX)

As reported in the *Washington Post*, the iPad has already been used to disseminate intelligence to US President Obama, "allowing analysts to add video and audio clips and interactive graphics."[3] The article quotes Shawn Turner, Director of Public Affairs for the Office of the Director of National Intelligence, describing tablets as a proper and secure alternative way to provide intelligence and likely be used more frequently in the future to represent multimedia information in the PDB.[4]

Although it is the responsibility of each nation's intelligence community to establish its own communication standards for analytic products, the very purpose of the intelligence is to deliver the best possible analyses in a timely and usable way to facilitate the decision making of the policymaker or consumer. Intelligence products are useful only if they provide information and analytic insights on time to the decision maker, reducing their uncertainty and the tension that it produces while

DOI: 10.1057/9781137523792.0006

facilitating decisions and posterior actions. Independent of whether the intelligence product is conceived to be delivered in a printed or in a digital and interactive format, the product needs to be delivered to the user in a timely fashion.

In principle, immediacy is one of the advantages of digital communication compared to printed media. In the absence of the necessary skills to design information (or intelligence), combine and integrate media in a meaningful way, and anticipate sequences of interaction by the user, however, the results can be counter-productive. Although the traditional principles of analytic writing remain essential, a core difference exists between textual reports and digital intelligence products:

> All traditional text, whether in printed form or in computer files, is sequential, meaning that there is a single linear sequence defining the order in which the text is to be read [...] Hypertext is nonsequential; there is no single order that determines the sequence in which the text is to be read [...] Hypertext presents several different options to the readers, and the individual reader determines which of them to follow at the time of reading the text. This means that the author of the text has set up a number of alternatives for readers to explore rather than a single stream of information [...] hypertext consist of interlinked pieces of text (or other information).[5]

A "digital turn" in the field of intelligence communication needs to take into consideration the fields of information design and interaction design, as well at the concept of user experience (UX). All of them provide a framework that sets the stage for digital communication and design for interactive media that can be useful for intelligence communication in the 21st Century.

As noted by Hartson and Pyla, the concepts of UX and design do not necessarily entail high-tech artifacts; technology is rather a design context.[6] Similarly, the concept of usability is critical. The user experience can be influenced by perceptions of the producing organization and past experiences. According to Nielsen, usefulness, defined as the capability of a system to be used to achieve a goal, can be broken down into two categories: usability and utility. Specifically:

> Utility is the question of whether the functionality of the system can do what is needed, and usability is the question of how well users can use the functionality.[7]

The concept of usability applies to all aspects related to the systems with which we interact. It consists of five attributes: learnability (easy

DOI: 10.1057/9781137523792.0006

to learn), efficiency (efficient to use), memorability (easy to remember), errors (low error rate), and satisfaction (subjectively pleasant to use).[8]

Accordingly, intelligence products should be designed by taking usability and its components into account. Figure 1.2 applies these usability attributes to the interaction of the intelligence consumer/user with analytic products.

In the field of interaction design, UX integrates the concepts of utility and usability but adds other components. UX is an expansion of the concept of usability design, entailing also "social and cultural interaction, value-sensitive design, and emotional impact – how the interaction experience includes joy of use, fun, and aesthetics."[9] UX is defined as:

> the totality of the effect or effects felt by a user as a result of interaction with, and the usage context of, a system, device, or product, including the influence of usability, usefulness, and emotional impact during interaction, and savoring the memory after interaction. Interaction with is broad and embraces seeing, touching, and thinking about the system or product, including admiring it and its presentation before any physical interaction.[10]

Adopting a user-focused approach when producing analyses is critical in order to be relevant for intelligence consumers. The concept of Intelligence UX highlights this necessity by considering not only the

Usability Attributes	Meaning	Application to Analytic Products
Learnability	Easy to learn	The structural organization (inverted pyramid approach) of the product facilitates interaction by the user. Degrees of uncertainty and the quality of sourcing are expressed using an easy system of words/numbers.
Efficiency	Efficient to use	The paper present a clear picture addressing the "so what" putting the bottom line up front, with key judgments and implications highlighted.
Memorability	Easy to remember	Layout template is easy to remember (title, headings, bottom-line, key judgments) and the paper tells a compelling story.
Errors	Low error rate	Avoidance of misspellings, grammatical errors, unfounded assumptions, and poor logic.
Satisfaction	Pleasantly used	Preference for one system (analytic product) over others, visually effective, attractive layout, and good use of graphics.

FIGURE 1.2 *Application of usability to traditional analytic products*

DOI: 10.1057/9781137523792.0006

most fundamental aspect of the utility of intelligence products, but also how it is related with their usability through a proper interaction design (of the user with the product). The design should take into account that satisfactory use (joy of use) by the client will have an impact in his/her willingness to use the system for making decisions.

Products that provide a satisfactory UX are more likely to impact decisions. In a world of information and cognitive overload, the analyst has to struggle to capture the attention of the intelligence client. It is not enough, although desirable and certainly the most important for the intelligence service mission, to collect the best possible information and provide the best possible analysis. Much more than in the past, intelligence analyses competitiveness is now affected by the manner in which an insightful analysis and strategic information is conveyed to consumers.

Although it seems obvious, it is important to highlight that in an environment of multi-touch screens where digital communication is the norm, organizations that only deliver printed products will be perceived as an exception. The expectation will be that high quality products must include digital input. In addition, narratives from the entertainment industry and images of high-tech intelligence services affect the cultural environment influencing the policymaker. In the corporate world, it will become unthinkable to provide competitive and market intelligence deliverables that are devoid of graphics and digital media.

Multimedia competencies and the intelligence analyst

The concepts of digital natives[11], net generation (1977–1997)[12], or generation C[13], among others, stress the implications derived from the impacts produced by the technological changes in the generations that have grown in the Digital Era. According to Palfrey and Gasser, digital natives:

> Were all born after 1980, when social digital technologies such as Usenet and bulletin board systems came online. They all have access to networked digital technologies. And they all have the skills to use those technologies.[14]

Although timeframes vary depending on sources, most authors agree that digital technologies have sociological, economic, psychological, political, and cultural consequences.

> They are Generation C – connected, communicating, content-centric, computerized, community-oriented, always clicking. As a rule, they were

DOI: 10.1057/9781137523792.0006

> born after 1990 and lived their adolescent years after 2000. In the developed world, Generation C encompasses everyone in this age group; in the BRIC countries, they are primarily urban and suburban. By 2020, they will make up 40 percent of the population in the United States, Europe, and the BRIC countries, and 10 percent in the rest of the world – and by then, they will constitute the largest group of consumers worldwide.[15]

Thus, the trend is toward a massive use of digital technologies both for the next and current generations. However, being able to use digital technologies and consume products in digital formats is quite different than being skilled at producing contents in digital formats. Education for providing digital literacy and training digital communication skills is required for being able to use digital technologies from the perspective of a producer. The author's experience teaching multimedia communication courses to students of journalism for several years as well as designing and conducting with colleagues the *Multimedia Intelligence Product* simulation exercise as part of a Master's Degree program in intelligence analysis for five years shows that being a user of digital communication devices and information does not equate with being a skillful producer.[16] In our classes, some Generation C users experience paralysis, spurred by their fear of technology, and are unable to perform clear step-by-step instructions when asked to build rather than to use digital devices. Getting familiar with a specific multimedia communication-related language and tools can pose additional challenges, particularly if the task involves computer programming.

Overcoming the initial frustration that results from not being able to successfully complete a task, understanding the reasons behind recurrent unsuccessful attempts, as well as activities such as successfully publishing for the first time a website or updating the version of a Content Management System are milestones that appear like rites of passage. The practice of digital journalism requires professionals with specific knowledge and skills. Higher education programs and continuing learning courses provide education and training in digital communication competencies. A digital turn in intelligence analysis communication requires practically the same competencies of digital journalism and specific training that stresses the existing differences in both practices.

Usability attributes also apply to the producer. Utility and usability are mandatory. The usefulness of an intelligence product from the perspective of the producer depends on how much additional value the system provides for better achieving the organization's mission and creating informative value for facilitating the decision making process. Insightfulness and timeliness are key. The system has to be usable and facilitate timely and satisfactory experiences. Timeliness is a priority. Back-end technology needs to be usable and to facilitate an optimal workflow. The number and complexity of tasks to perform by the analysts in order to convey the best possible analyses in multimedia formats has to be not higher than the ones to perform when writing analytic papers in text-only format. A mockup of a graphical user interface model for guiding analyst through the process of writing the intelligence reports, including help tooltips, and basic reminders on the principles of analytic writing was presented by the author at the SCIP European Summit 2013.[17]

Wireframing an intelligence report

How does a multimedia report look? The answer to the question depends on the nature of the intelligence product. The conceptualization and design of the products will depend on whether the analysis is of a basic, current, estimative, or indications and warning nature. It will also depend on the mission and culture of the organization and should be informed by consumer feedback.

Based upon the structure of an unclassified US National Intelligence Council report, Figure 1.3 sketches the information structure, labeling, and interaction of a hypothetical digital intelligence report.[18] This wireframe of the multimedia report is similar to one proposed to students as a template for running the simulation exercise previously cited. The participants produce these multimedia reports as part of their training.

Specific products of other intelligence communities like unclassified UK Joint Intelligence Committee (JIC) Assessments can also inspire the structure of the reports. The static representation of the interactive product involves navigation through the following tabs and buttons:

- Scope Note
- Bottom-line
- Key Judgments
- Background
- Assumptions

DOI: 10.1057/9781137523792.0006

- Video-briefing
- Scenarios
- Sourcing and references (including source evaluation)
- Export as E-paper (an embedded intelligence report that can be exported as in portable document format)

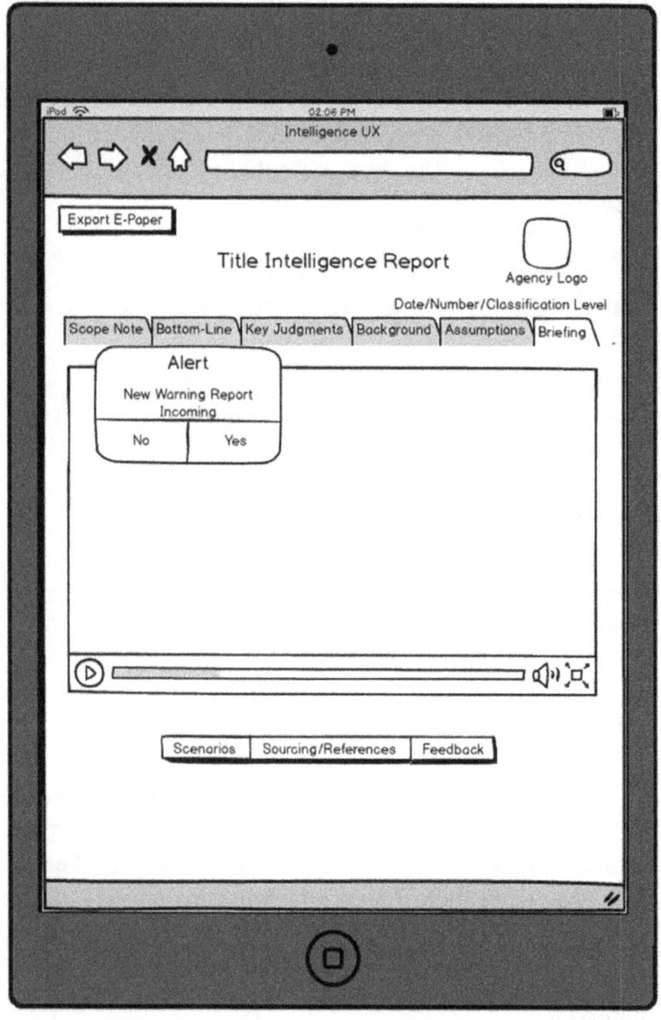

FIGURE 1.3 *Sketchy wireframe of multimedia report created with Balsamiq mockups*

Source: Ruben Arcos, 2015

DOI: 10.1057/9781137523792.0006

Given the nonsequential nature of digital products, it is important to establish a product map and anticipate scenarios and paths for the user. Labeling should be consistent with the content and support easy navigation. An opportunity provided by digital communication is that oversight bodies or internal reviewers can introduce additional requirements for evaluating the correctness of the analysis and make more transparent the analytic process allowing its evaluation. Digital products can help in tracking analytic bias.

A variety of multimedia elements can assist in providing valuable intelligence products to consumers. The question is not how many of them to include but which are most appropriate. Interactive infographics, maps, timelines, static image sliders, audio, customizable and interactive charting, diagrams, network visualization, and explanatory videos can be integrated in the report. Many of these tools and solutions can already be found in the commercial market. The business community uses several well-known examples of competitive intelligence proprietary software.

A problem that usually surfaces with the use of this software is the lack of an appropriate information design and internal organization of the reports. In other words, analysis has to be provided through a carefully structured presentation of the key judgments and findings and supported by a solid argumentation. It is not just facts. It is about extracting and explaining meanings and implications, as well as anticipating developments.

Garrett's conceptual framework for designing user experience in the case of Websites provides a clear model that can be useful for conceptualizing multimedia intelligence reports.[19] According to Garrett's framework, five planes affect the composition of a website: strategy, scope, structure, skeleton, and surface. Each plane is dependent on the planes below it (see Figure 1.4).

- The **strategy plane** considers the expectations of the user about the product as well as the organization objectives in relation to the product and those needs.
- The **scope plane** translates the strategy plane into specific requirements "for what content and functionality the product will offer to the users."[20] Content requirements refer both to text and multimedia elements like the ones above mentioned, but it is necessary to remember that the purpose of the content is the primary concern. Format should not make us forget the informative and cognitive value of the analysis.

DOI: 10.1057/9781137523792.0006

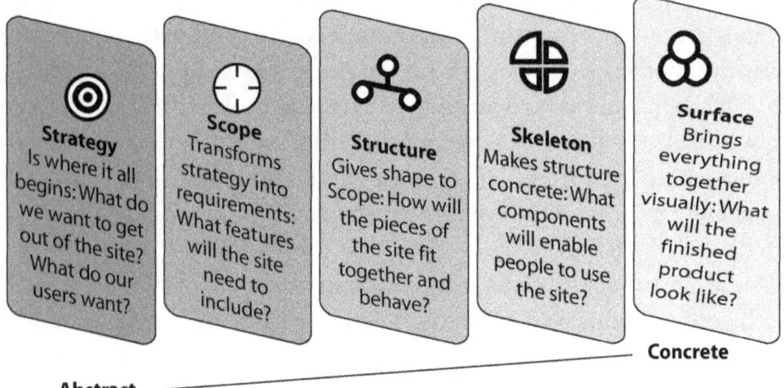

FIGURE 1.4 *The five planes of the user experience*

- Garrett links the **structure plane**, which brings concretion to the product, to interaction design and information architecture. The first has to do with how the user will interact with the systems and how the system will react to the user behavior. Information architecture is concerned with how contents are structured for facilitating understanding and use ranging from hierarchical to free exploration organic sites.
- The **skeleton plane** depends on navigation design, information design, and the design of the interface. Wireframes connect the three of them in a schematic design. According to Treder, wireframes are low fidelity depictions of a design showing core groups of content, the structure of information, and basic visualizations of actions between users and the interface, a kind of backbone of the design, which is consistent with Garrett's concept of the skeleton plane.[21]
- The **surface plane** deals with concreting the visual design and what Garrett names sensory design. This includes color palettes, typography, and branding elements like the logo of the organization.[22]

Structured briefings and face-to-face interaction

Multimedia intelligence products and digital communications should not be considered as substitutes for face-to-face interaction with the

DOI: 10.1057/9781137523792.0006

decision maker. Rather, multimedia and digital communications are aids supporting the explanatory function of intelligence analysis. Interpersonal interaction is key to building trust and emotional involvement. Technology in this context is conceived to support understanding and communication, not to increase the distance with the client. For instance, if the client is too senior or too busy, a video briefing by an experienced analyst is better than a no-briefing scenario.

Face-to-face presentations require specific skills and preparation. Deep knowledge and experience builds trust but does not guarantee a persuasive and compelling discourse or a superior speaking performance.

Immersive communication and augmented reality open new ways to interact with the customers and provide briefings. Again, bringing structure to the content under a proper timeframe is key. Digital communication provides opportunities to increase the number of presentations. Digital communication and augmented reality will compel analysts and their managers to rethink the portfolio of intelligence deliverables and presentations aimed at supporting the decisions of the C-Suite. In the case of intelligence services and classified information, digital communications can be challenging to information security policies and standards. However, the need to adapt – sooner or later – hardcopy, narrative analytic products to the digital era will be necessary to remain competitive. It is the norm to find high levels of resistance to change within intelligence organizations with a toughly rooted secrecy culture. In these cases, it is important to remind them – as observed by Berkowitz and Goodman – that secrecy should be considered a tool of the trade; one that imposes costs and that should be managed intelligently.[23]

Notes

1 This chapter is based on the paper "Producing and Consuming Intelligence Products in the Digital Era: The Need for Multimedia Communications," prepared by the author for presentation at the panel "Reinventing Intelligence Production for the 21st Century," International Studies Association, Toronto, Ontario, Canada, March 29, 2014.

2 Vic Costello with Susan A. Youngblood and Norman E. Youngblood, *Multimedia Foundations: Core Concepts for Digital Design* (Waltham, MA: Elsevier, 2012), p. 12.

DOI: 10.1057/9781137523792.0006

3 See Greg Miller (2012). "Oval Office iPad: President's Daily Intelligence Brief Goes High-Tech," *Checkpoint Washington*; http://www.washingtonpost.com/blogs/checkpoint-washington/post/oval-office-ipad-presidents-daily-intelligence-brief-goes-high-tech/2012/04/12/gIQAVaLEDT_blog.html.

4 An official photo (dated January 31, 2012) of the President Barack Obama using a tablet computer while receiving the Presidential Daily Briefing can be consulted at the following link: http://www.whitehouse.gov/photos-and-video/photo/2012/01/president-barack-obama-receives-presidential-daily-briefing.

5 Jakob Nielsen, *Multimedia and Hypertext: The Internet and Beyond* (Mountain View: Morgan Kaufmann, 1995), pp. 1–2.

6 Rex Hartson and Partha S. Pyla, *The UX Book: Process and Guidelines for Ensuring a Quality User Experience* (Waltham, MA: Morgan Kaufmann, 2012).

7 Jakob Nielsen, *Usability Engineering* (Mountain View: Morgan Kaufmann, 1993), p. 25.

8 *Ibid.*, p. 25.

9 Rex Hartson and Partha S. Pyla, *The UX Book: Process and Guidelines for Ensuring a Quality User Experience* (Waltham, MA: Morgan Kaufmann, 2012), p. xii.

10 *Ibid.*, p. 19.

11 Prensky, Mark (2001). "Digital Natives, Digital Immigrants", *On the Horizon*, Vol. 9 (5); http://marcprensky.com/articles-in-publications/.

12 Don Tapscott, *Grown Up Digital: How the Next Generation Is Changing Your World* (McGraw-Hill, 2009).

13 Roman Friederich; Michael Peterson; Alex Koster and Sebastian Blum, *The Rise of Generation C: Implications for the World of 2020* (Booz & Company, 2010); http://www.booz.com/media/file/Rise_Of_Generation_C.pdf.

14 John Palfrey; Urs Gasser, *Born Digital: Understanding the First Generation of Digital Natives* (New York: Basic Books, 2008).

15 Roman Friederich; Michael Peterson; Alex Koster and Sebastian Blum (2010). *The Rise of Generation C: Implications for the World of 2020* (Booz & Company, 2010); http://www.booz.com/media/file/Rise_Of_Generation_C.pdf.

16 The simulation has been published in a ready-to-run format. See: Rubén Arcos; Manuel Gértrudix;, José Ignacio Prieto, "Multimedia Intelligence Products: Experiencing the Intelligence Production Process and Adding Layers of Information to Intelligence Reports," in William J. Lahneman and Rubén Arcos, eds, *The Art of Intelligence: Simulations, Exercises, and Games* (Lanham: Rowman & Littlefield Publishers, 2014), pp. 239–260. Participants have widely benefited from the expertise of professors Sergio Álvarez and Manuel Gértrudix in the field digital communication.

17 Rubén Arcos and Manuel Gértrudix, "Apply Multimedia Technology for Intelligence Reporting", presentation delivered at the European SCIP Summit 2013, Rome, November 5–7, 2013.

DOI: 10.1057/9781137523792.0006

18 Global Water Security. Intelligence Community Assessment, ICA 2012–08, February 2, 2012; Available at: http://www.dni.gov/files/documents/Special%20Report_ICA%20Global%20Water%20Security.pdf.
19 Jesse James Garrett, *The Elements of User Experience: User-Centered Design for the Web and Beyond*, 2nd edition (Berkeley: New Riders, 2011).
20 *Ibid.*, p. 60.
21 Marcin Treder, (2013). *UX Design for Startups*. Available at: www.uxpin.com.
22 Jesse James Garrett, *The Elements of User Experience: User-Centered Design for the Web and Beyond*, 2nd edition (Berkeley: New Riders, 2011), p. 156.
23 Bruce D. Berkowitz, and Allan E. Goodman, *Best Truth: Intelligence in the Information Age* (New Haven: Yale University Press, 2000), p. 160.

DOI: 10.1057/9781137523792.0006

2
Presentational Tradecraft: A New Skill

Mary O'Sullivan

Abstract: *Modern-day drafters of analysis must develop presentational tradecraft skills to meet customers' expectations that every product will have one or more accompanying visuals to supplement static narrative. Analysts must design a product with photos, maps, links, and interactive elements from the start and package the story for maximum relevance, retention, persuasion, and retrieval. The CREATE framework, which focuses on customer relevance and ease of use, can help analysts conceptualize products that take full advantage of digital communication. By applying some basic storytelling principles of design and persuasion to their presentational tradecraft while using the CREATE model, analysts can make analytic products more pleasing and more memorable to the customer.*

Keywords: design; layering information; persuasion; presentational tradecraft; storytelling

Arcos, Rubén and Randolph H. Pherson, eds. *Intelligence Communication in the Digital Era: Transforming Security, Defence and Business.* Basingstoke: Palgrave Macmillan, 2015. DOI: 10.1057/9781137523792.0007.

 DOI: 10.1057/9781137523792.0007

Like other modern-day users of information who have made the transition from the printed page to the web, most consumers of intelligence products now access finished analysis in electronic formats. The ideal product is "user-driven" or "layered" with material that explains, defines, or visually depicts what consumers are reading. Consumers can read as much or as little as they want, depending on their interests and needs.

The move to create multimedia presentations and to place intelligence analysis on mobile devices for senior US policymakers is a reflection of the power of digital communication as well as consumer's desires. The inclusion of "visualization" in an updated Intelligence Community Directive on analytic tradecraft standards issued by the US Director of National Intelligence is further evidence of increased emphasis on graphics, maps, still photos, video, or other technology to supplement data and static narrative analysis.

Asking analysts to include graphics, maps, and other visuals in finished analysis is not a new phenomenon. What has changed is consumers' expectations, stemming primarily from their personal web use. They expect that every analytic product will have an accompanying visual. Hiring scores of new graphic designers or cartographers or multimedia developers is not the answer to increasing demand, though some organizations will likely need to add some personnel with advanced design, editing, and web-content experience. Analysts themselves and their managers must develop skill in writing for the web as well as visualizing what image or media will drive home the analytic point of the product.

Presentational tradecraft[1]

Modern-day drafters of any type of analysis – strategic, tactical, or competitive business intelligence – must develop presentational tradecraft skills that focus on three elements:

- How an intelligence product looks – the images that help convey the meaning of the story as well as writing to accommodate typical web-reader behavior.
- How a piece of intelligence comes to the user – where it fits in the continuous stream of information available and the user's capacity to access the material.

DOI: 10.1057/9781137523792.0007

- How a piece of intelligence is organized or "unfolds" – with the "key analytic judgment" up front, supported with data and solid evidence.

Analysts must learn to plan a product with visuals, links, and interactive elements from the start, rather than adding illustrative material as an afterthought. Writing in "user-driven" rather than "user-absorption" formats and finding the right balance between too many and too few layers of information require that analysts focus on the customer and package the story for maximum relevance, retention, and retrieval.

The temptation of templates

Creating product templates that reflect web readers' behavior and that provide brand recognition would appear to be an easy solution to help analysts who were hired for their substantive expertise, not graphic or visual design skill, to deal with the changing product landscape. A format that analysts, designers, and cartographers simply "fill in" and that guides the reader to chunks of content is consistent with current practice across the US Intelligence Community. New designs could reflect eyetracking studies and web-usability research.

- The Poynter Institute has been pioneering EyeTrack research for journalists since 1990. Their studies reveal that most on-line readers now scan material in a "Z-like" pattern, focusing first on headlines and summaries, then turning to photos and graphics.[2] New product designs can easily reflect this reading behavior.
- Many intelligence agencies' increased focus on well-written summaries is consistent with eyetracking studies that show summaries are essential in digital communication. When participants in the most recent studies encountered an introductory paragraph, 95% of them read all or part of the paragraph. The downside is that readers who spent the most time with a story's introduction spent the smallest amount of time on the body of the story. Thus, summaries that are comprehensive have the most impact.
- Writing guidelines that exhort drafters to keep it short and simple are also on track with digital reader behavior. Limiting the amount of text in a paragraph encourages readers to continue reading.

DOI: 10.1057/9781137523792.0007

> Short paragraphs, limited to one or two sentences or one thought, are twice as "popular" as paragraphs with six or more sentences.

The danger of rigid template structures is that they discourage thinking before writing and thinking differently about the web-based product. Pointers or signposts that direct the reader to elements of the story are aids to comprehension, but studies show that pointers must have "information-carrying content," not simply describe what the reader should expect in a section or portion of a product.

- Writing is an essential part of the web-based product, but now analysts need to use precise words that add new layers of understanding, not repeat information, and think about the links that need to be included. A danger is to rely too much on new technology to carry the story and to become a sloppy writer limited by the number of "allowed" character strokes.[3]
- Choosing subtitles or section heads is also more important than ever when writing for the web. Jakob Nielsen, a web-usability expert, has characterized web users as engaged, active readers who want to go places and get things done. "Discussion" or "Outlook" – typical section headers currently found in many intelligence agencies' products – are not keywords that aid in comprehension or retrieval.[4]

The CREATE model

Pherson Associates has developed a simple mnemonic – CREATE – to help analysts conceptualize products that take full advantage of digital communication and assist with the transition from static text to more dynamic modes of content delivery (see Figure 2.1). Applying some basic principles of design and persuasion while using the CREATE model can make analytic deliverables more pleasing and more memorable to the customer.

Like the AIMS – Audience, Intelligence Question, Message, and Storyline – model that Pherson Associates has incorporated into its training materials, CREATE prompts analysts to consider the elements of presentational tradecraft. It contains some of the same concepts as AIMS but elaborates on them and adds a few more.

Customer refers to the need to have a specific audience in mind for every analytic product. The customer can be the original requester of a

DOI: 10.1057/9781137523792.0007

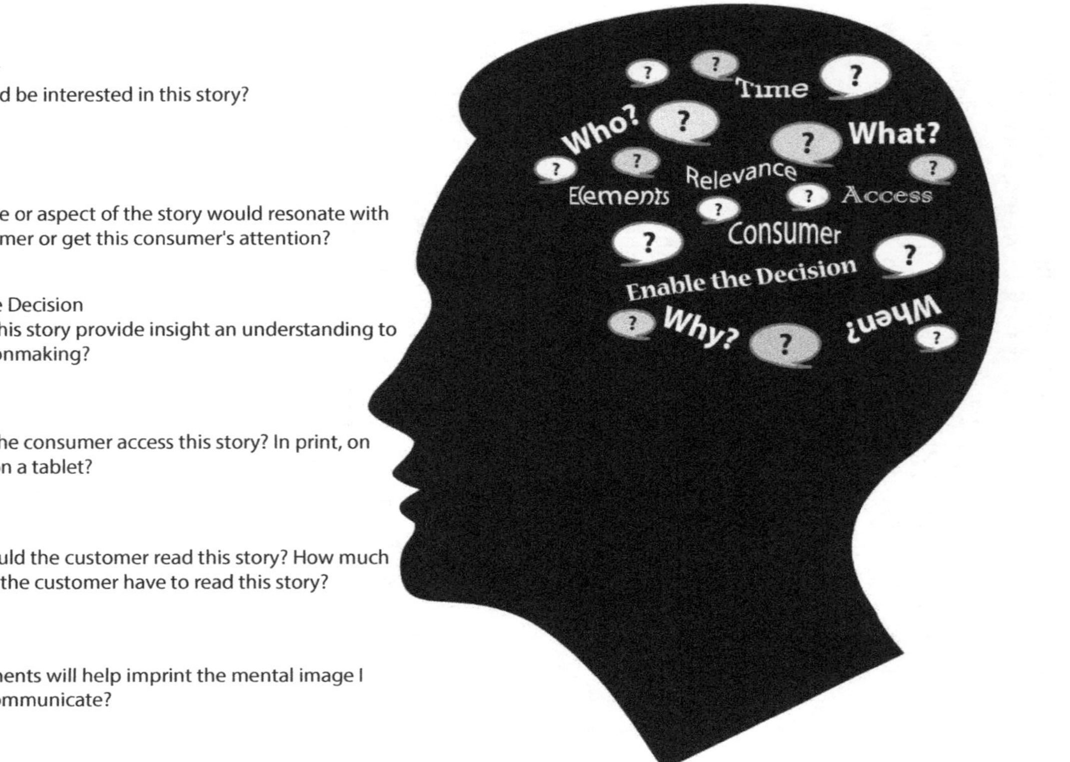

FIGURE 2.1 *CREATE – a guiding framework*

Source: Pherson Associates, LLC, 2015

DOI: 10.1057/9781137523792.0007

product, but there may be many others who would be interested in the topic. Posing a few simple questions can help decide whether one needs to write a short, tactical piece or a less-detailed strategic piece or both. The questions are:

- Who cares about this topic or issue or story? Are there new as well as returning customers? Is this an update to an ongoing story? If so, what do customers know thus far and where have they obtained their information?

Relevance refers to the interests of various consumers. Are all the customers who care about the topic interested in the same aspect of the topic? In a web-based production environment, analysts can accommodate the interests of more and varied customers than in a static or print format.

Mario Garcia, a Poynter Institute fellow who has designed many on-line newspapers asks this question: "How can I accommodate the quick read, the substantial read, and the encyclopedic read?"[5] Layering information will accommodate multiple types of readers. Layering is not simply adding links here and there. A layer of information must stand alone, as if someone stumbled across it and could get the gist of what is going on by reading only that piece of information.

Enable the Decision refers to providing the consumer with perspectives that add insight or give the consumer the context needed to understand the So What? or the Why? explanation contained in the analysis. Such questions might include: What are the levers of power that the customer(s) has at his/her disposal? How is the development or story unfolding? What are the key decision points? Who are the dominant decision makers on this issue?

Access refers to knowing how customers read. Retrieval is a main consideration when writing for other analysts who access material on the web and who might likely use it in their product or save it for later reading. Including words in the common lexicon of customers who follow the issue is essential. Also, it is more difficult to retrieve content from video or audio material than from text.[6] Is a staffer who prepares briefing books a likely consumer of the product? How would one package material for a senior consumer who has a designated briefer? Does the senior like reading on a tablet?

DOI: 10.1057/9781137523792.0007

- Research into how people read on tablets shows that text actually works quite well because tablet users read longer on the device than folks who are working on a laptop. Also, tablet readers spend more time reading material on a tablet in the evening. Though current security technology may preclude senior government officials from using tablets at home, secure Wi-Fi will surely come soon.[7]

Time has two dimensions: when should the product arrive in the hands of the consumer and how long will the consumer likely spend on the story? Creating multimedia products is generally time-consuming; until an analyst has developed a "library" of various kinds of original media, each multimedia project will be labor-intensive. The inclusion of readily available media, i.e. social media, news sources, and so on is a possible solution when following a fast-breaking story. Adherence to good tradecraft requires that the source be properly caveated and the reader informed regarding the potential for alteration.

Assessing how long the customer will be interested in the story either from an issue perspective or from a scheduling perspective is guesswork. In the case of senior executives, calling the briefer, staffer, or others to ascertain the interests of, patience for, and schedule of the primary consumers is essential if one hopes to place a complicated digital technology story in his or her hands.

Elements refer to the type of visuals you will use to convey your story. The first and fundamental question is: does this story lend itself to a multimedia presentation? Will video or audio or animation add something to this story? What are you writing about: a trend, a process, a new discovery? Are you comparing old with new? Would a simple pie chart convey the intended message if time is of the essence?

Focus on the producer and the user

Relevance – the second factor in the CREATE model – dictates that the customer or target audience be "king." In the world of digital communication, there are many kings, queens, and even squires! How is the analyst-producer to know who will read the product?

Rubén Arcos makes the case that creators of web-based products – software developers, analysts, and graphic designers – should look to the field of interactive design and adopt the concept of User Experience (UX), a term that encompasses the ease of use, utility, and socio-cultural

DOI: 10.1057/9781137523792.0007

aspects of joy, fun, and aesthetics when fashioning a product. Designing a pleasing, consistent, and easy-to-use web-based product is nothing new. Few customers will return to a brand that has too many bells and whistles, is inconsistent, or is hard to use. Arcos points out that UX should now encompass both consumer AND producer. The number and complexity of analytic tasks associated with digital technology should not exceed the requirements for text-only formats. Few analysts will be eager to create new multimedia products if they result in an exponential increase in workload.[8]

Analysts have been accustomed to creating multiple versions of an intelligence story to accommodate customers' interests and "need to know." Layered products, however, are a far greater change than past requirements, such as "writing for discovery," which asked analysts to segregate content above and below tear lines according to the clearances and interests of potential readers. Analysts must now accommodate the "quick reader, the substantial reader, and the encyclopedic reader" in one product.[9]

As already noted, product design based on eyetracking studies can assist the "quick reader" who wants to scan but not scroll. A long-held principle of analytic writing, the Inverted Pyramid – with the key analytic judgment as lead followed by the 2–3 most important points – will give the substantial reader the information he or she wants or needs without further navigation. Linking or layering will accommodate the encyclopedic reader but: What to link? How much to link? Where to place links? How links should relate to one another? Jakob Nielsen says that links should be "a set of pyramids floating in cyberspace."[10]

The reality is that many analysts are more likely to create the Leaning Tower of Pisa or the Tower of Babel rather than a nicely-architected building or home. Serving on organizational task forces that generate SitReps (Situation Reports) twice or thrice a day exacerbates the tendency to "Pile-On" or simply add the latest development in a fast-breaking story on top of existing material.[11] Such SitReps make sense to those who are following the story or issue closely but often do not stand the test of time.[12]

Creating "archival links" that contain enduring foundational readings, record timelines of important developments, state key analytic assumptions, and posit indicators of change may help overcome the "Pile-On" effect. Such links can yield two benefits: the encyclopedic reader can find needed or desired detail that is contextually anchored, and the analyst

DOI: 10.1057/9781137523792.0007

has the opportunity to create longer, in-depth think pieces that are structured using digital technology. Moreover, users far prefer an easy-to-navigate searchable framework of understanding rather than links to pages that then require further scrolling and clicking.

Storytelling for "stickiness"

Storytelling has been part of the human tradition since the caveman. Telling a compelling story is the key to the third element of presentational tradecraft – how the story unfolds so that it resonates with the reader and is memorable. Storytelling is also an essential principle of persuasion or "stickiness," later discussed in this chapter.

Every intelligence product is actually telling a story: a story of threat – acquisition of new weapons; a story of betrayal – violation of economic sanctions; or a story of reassurance – the peaceful transition of power from one leader to the next. The following table compares the essential elements of storytelling with intelligence analysis, which are closely related to the Who, What, How, When, Where, Why, and So What of journalism (see Figure 2.2).

Plot is the events that make up a story. Intelligence analysis centers around events – things that have happened, will happen, or could happen. Strong intelligence stories capture WHAT has, will, or could occur and **HOW** the event(s) described or projected came or could come about.

Every story has **Characters** – sometimes they are heroes, sometimes demons, sometimes tragic figures. There are characters or actors in every topic of interest in the intelligence world; sometimes the characters are actually organizations. We often ask: **WHO** is pulling the strings? What organization is the power broker? How is this or that group organized?

Intelligence Analysis	Storytelling
What/How	Plot
Who	Characters
When/Where	Setting
Why	Theme
So What	Goal

FIGURE 2.2 *Contrasting the essential elements of intelligence analysis and storytelling*

Source: Pherson Associates, LLC, 2015

DOI: 10.1057/9781137523792.0007

Everything happens at some point on the globe or in the atmosphere – that is the **Setting** or sometimes the context of **WHEN** or **WHERE** something occurred or might occur that helps put things in perspective. Many times the "background" element of a product contains the essential ingredients of the setting.

The **Problem** is the explanation of **WHY** something is happening. In a novel, the problem is the issue with which the protagonist is dealing; the issue explains why the characters do what they do. Sometimes fiction writers use themes such as friendship or betrayal. In intelligence analysis, the problem or theme is stated most often in the message.

- Country X's troop movement toward the border is a military exercise likely designed to intimidate its country's neighbors. You are reporting because the troop movement is both the problem and the event you are addressing.
- Country X has passed a critical milestone in the development of a new missile system capable of hitting the US homeland. This new missile system is the problem you are reporting on. The event is the milestone.

The **Goal** is the **SO WHAT** or the implications of an event. In fiction, the problem and the goal are often closely related. In intelligence analysis, we posit that: if what we say is correct, then, carried out to its fullest extent, this is likely to happen.

- The XXX militant group is advancing on the country's capital. If its rate of advance continues at its current pace, the group will be in the capital's suburbs within XXX days.

Storytelling engages readers by relating the analyst's argument to the readers' experiences, by framing an argument in terms familiar to the reader either through culture or training, by selecting a visual that brings the issue to "life." Digital communication offers the analyst the opportunity to visualize the picture and organize the elements of the story in narrative form. Multimedia can help a story grip the reader's imagination: the potential spread of a disease is best told through animation rather than text; the death of a coral reef told in photographs accompanied by the lonely sound of gulls is far more powerful and no less legitimate than tables measuring the branch-by-branch destruction of an endangered ecosystem.

Many analysts are troubled when urged to "tell a story" because they have been trained to think critically, systematically, and logically.

DOI: 10.1057/9781137523792.0007

Thinking logically and storytelling are not at opposite ends of the spectrum, however. Both are legitimate ways of organizing information, with the former relying primarily on the power of argument and the latter relying on analogy and "lifelikeness" to make the case. In every instance, the analyst must be aware of his or her own assumptions and biases, avoid logical fallacies, and misused emotion.

Digital communication, persuasion, and "stickiness"

In the book *Made to Stick*, Chip and Dan Heath explored the idea of why some ideas or stories endure while other equally worthwhile ideas wither and die.[13] They found that the ideas that endured shared six different characteristics. They were:

- Simple – contained a single, replicable thought.
- Unexpected – generated curiosity and interest.
- Concrete – were anchored in human interactions or things that humans can comprehend.
- Credentialed – presented by knowledgeable experts or able to be tested easily by the layman.
- Emotion-filled – generated feelings.
- Story-based – presented in stories.

Analysts who integrate maps, video, photos, audio, and animation into a text narrative are using the power of digital communication to increase the chances that their ideas will stick! Not all stories lend themselves to each of the aforementioned methods of communication, and the inclusion of multimedia that adds little to the story can annoy the reader. Nonetheless, taking the time to consider what message you are trying to convey and what "image" you want to leave in the mind of the reader is worth the time.

The first step in thinking about the visuals to help cement a story in the minds of readers is to ask what the goal of the product is. Is the product demonstrating a new process? Asking readers to recall facts? Showing an upward or downward trend?

- Poynter Institute studies reveal that readers recall the steps in a process or procedure better when the material is presented in multimedia formats. Animated graphics are effective for identifying new terms or concepts and then demonstrating their application.

DOI: 10.1057/9781137523792.0007

- Text is most effective for recalling facts, such as names, dates, or places, though thumbnails or timelines are effective supplements to the text.[14]

The next step is to consider what persuasive techniques or principles of "stickiness" will best resonate with the reader. Selecting persuasive techniques may give some analysts pause, particularly if they equate persuasion with manipulation or advertising campaigns.[15] At its heart, however, all analytic writing is persuasive writing. Every drafter has a point of view and the purpose of the analytic product is to present that point of view in a convincing way – not to dupe or control someone, which is the purpose of manipulation, but to inform the reader's thinking, provide a new or perhaps different perspective, or to alert or update readers about developments that are consistent with or that represent the need for a change of strategy or operations.

Each of the principles of "stickiness" is applicable to analytic writing presented in digital form. The inclusion of too many persuasive principles, like the use of too many different kinds of visuals just because you can, will overwhelm the reader. Thoughtful selection is the key.

Simplicity argues for a "primary" visual that is the essence of the maxim: "A picture is worth 10,000 words." An excellent illustration of this principle is the opening of *A Game of Shark and Minnow, The New York Times* 2013 multimedia news graphic that documented how the Philippines and China were staking their claims to contested territory in the South China Sea. The lone sailor, the small boat, the expanse of sea – all conveyed the disparity first captured in the title.[16]

Unexpected can mean a title that draws in the reader, not in sensational fashion, but in a way that helps the reader understand why he or she should read further. *Accordions Are Not Always The Answer for Complex Content* is a catchy and unexpected title from a May 2014 article on the Nielsen Norman Group website.[17] (In web design, an accordion menu is a vertically-stacked list of headers.) The reader unfamiliar with the term is certainly intrigued by the connection between a musical instrument and complex topics.

Concrete is the principle behind analogies or contextual images that help frame the problem for the reader. A map of a distant or foreign flooded area superimposed over the reader's equivalent state, province, or region provides the scope of devastation in an easily accessible way.

DOI: 10.1057/9781137523792.0007

Credentialed is the "go to" or most easily comprehended principle of persuasion for most analysts. The inclusion of "expert testimony" or sourcing, reliability, and confidence statements are explicit manifestations of reasons to believe the information or conclusion in the analysis. Video or audio substantiated by "experts" – such as clips of terrorist leaders with accompanying analysis of the clip's veracity, time, and place – is a powerful persuasive tool.

Emotion-filled does not mean tugging at the heart strings of the reader but means using visuals that evoke a response of some type. An interactive graphic showing the spread of a virulent disease is emotion-based as it can spur the reader to provide humanitarian aid or to take preventive health measures; similarly, an animated graphic that depicts the world "going dark" because of the destruction of a communications satellite is likely a more powerful call to action than a static table documenting what communications networks ride on what satellites.

Story-based means organizing the information in a way in which the reader can personally relate. The *London Guardian's* use of a digital clock contrasting the profit made by garment companies and the wages paid to garment workers in Bangladesh during the time the reader spent on *The Shirt On Your Back* – a 2014 interactive documentary – was a way of personally involving even the casual reader who did not go on to read the entire product.[18]

Simple principles of design

Novices at presentational tradecraft should know four basic principles of design that are relevant to digital communication. Having some idea of what visuals to include and where to place them will facilitate discussions with graphic designers and multimedia developers and theoretically speed up the production process. The principles outlined below certainly will seem like common sense to experienced producers of analytic work.

In the *Non-Designer's Design Book*,[19] Robin Williams argues that every well-designed piece of work, including church bulletins, organizational newsletters, and even invitations, should reflect four basic principles of design: proximity, alignment, contrast, and repetition (see Figure 2.3). Most analysts inherently know that **Proximity** or grouping things together is a good way of organizing information. For example, an analyst

DOI: 10.1057/9781137523792.0007

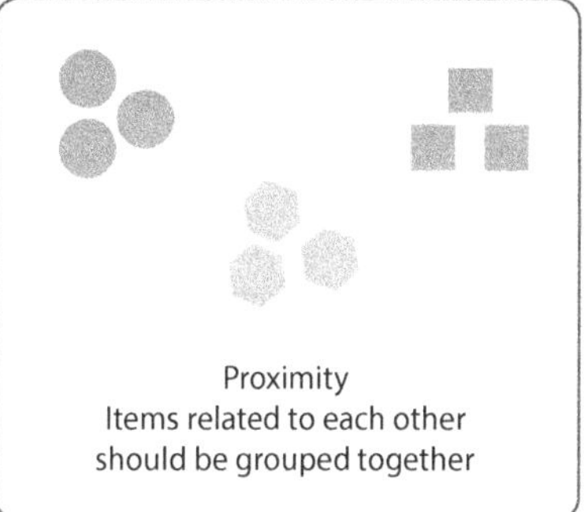

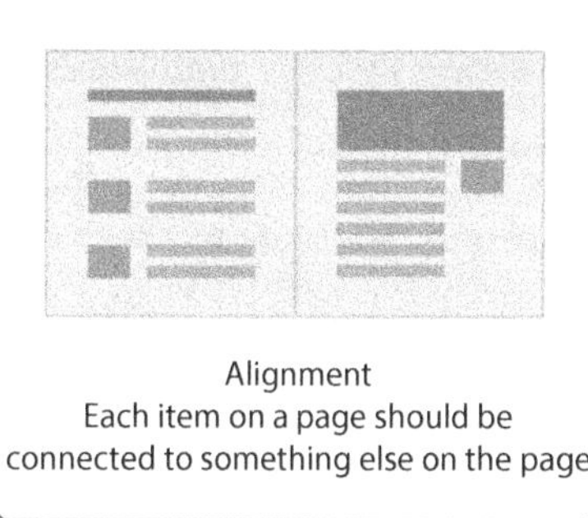

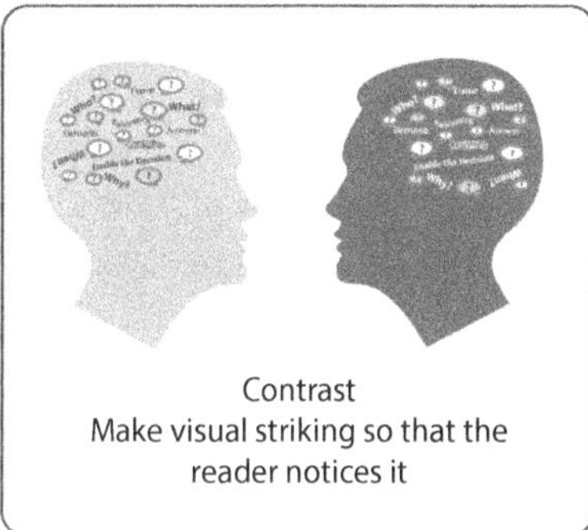

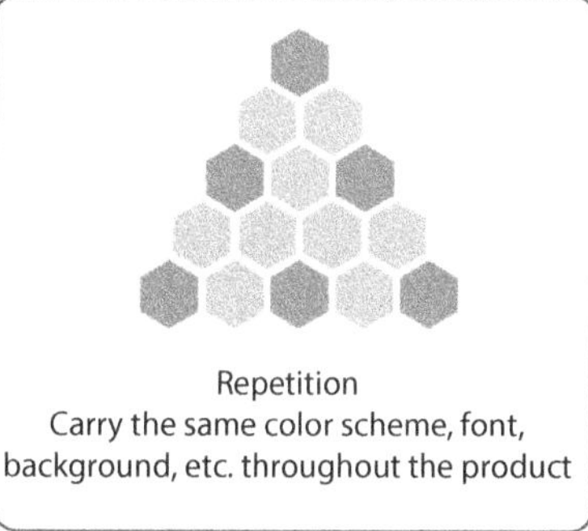

FIGURE 2.3 *The four basic principles of design*

writing a product about the six launch sites associated with a specific missile system would likely include the major shared characteristics of the launch sites in the summary but treat each launch site separately in the body of the paper. Each launch site would be treated comprehensively before moving on to the next. An alternative way of organizing the information would be to discuss each of the shared characteristics of the launch site system in a distinct and separate section. However, organizing the information that way makes it hard for the reader to get a "picture" of each site in his or her mind. An accompanying chart would likely list all the launch sites on one axis and use the other axis to record important characteristics, such as number of launchers, number of equipment trailers, and so on, at the sites.

Proximity also refers to the placement of graphics or images. Readers are more likely to look at a chart or graphic if it is close to the explanatory text. Though it may be easier to place all images, graphics, or maps in one section of a product, asking readers to toggle back and forth or scroll

DOI: 10.1057/9781137523792.0007

up and down is not a recipe for readability. Imagine a product discussing the leadership of an organization and the relationships among individuals in that organization but the product places all personality photos at the end of the text and after the graphics showing linkages among them. Thumbnail portraits of each person when the person is first named in the product would be a more effective way of cementing the individual in the reader's mind.

Alignment is also about organizing and unifying information or ideas by visually connecting with other items on the page or in the design. Even non-designers inherently know this principle. For example, when a drafter selects photos or images for a product, he or she typically will put all horizontally-oriented images together and place vertically-oriented images together. Also, captions or headers are consistently placed within a given category. For example, all captions are consistently at the top or bottom of photos, not oriented on the photo's axis. (Imagine the poor reader craning his or her neck from side to side as he or she attempts to read the photo captions!) The same principle of alignment applies to charts or matrices. Chart labels should be at the same place on each chart.

The principles of proximity and alignment both pertain when creating infographics. Text boxes should be either nearest the infographic feature they explain or in a designated section. Mixing and matching – some text boxes here, photos there, and other text boxes placed for typeface accommodation – is a recipe for confusion.

Contrast is about interest AND organization. If items discussed in a product are different in the analyst's view, then the items should stand out and be obviously different in visual treatments. For example, in a paper on different missile systems, readers should "see" immediately the size or throw weight or range differences among systems. An effective way of showing growth or diminution of a building or even an organization is to superimpose a line drawing on an image of the building or a selected representation of the organization. For example, the size of a parliament under a new constitution could be contrasted with the size of the former parliament by showing "doll figures" of two different colors on top of a picture of the parliament building or another well-known legislative edifice.

As just mentioned, color is a natural way of contrasting elements. Though an analyst may believe he or she has little sense of color palette and prefers to leave the specific choice of colors to a graphic designer, color contrast is everywhere. On maps, oceans are typically blue because

DOI: 10.1057/9781137523792.0007

they represent how our mind's eye "sees" the ocean and because cartographers need to differentiate water from land masses. Analysts should take care in using color, however, because various societies react to colors differently.[20]

Repetition is a reinforcing principle for a product. Choosing a design theme unifies the product and concept and adds visual interest. Choosing a strong design element – particularly one with which customers can identify is a winning idea. For example, analysts inherently know when commissioning a series of graphics about various missile systems that keeping the same design for missile X throughout the graphic series is essential.

Naming conventions are another manifestation of the principle of repetition. Intelligence analysts agree on a preliminary name for a newly-spotted ship or aircraft and all agree to call it XXXX until it becomes clearer exactly what type of ship or aircraft it is. All readers look for consistency and the principle of repetition lends consistency.

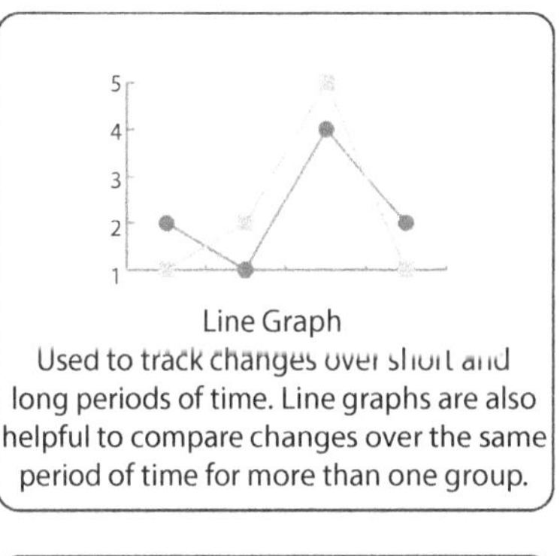

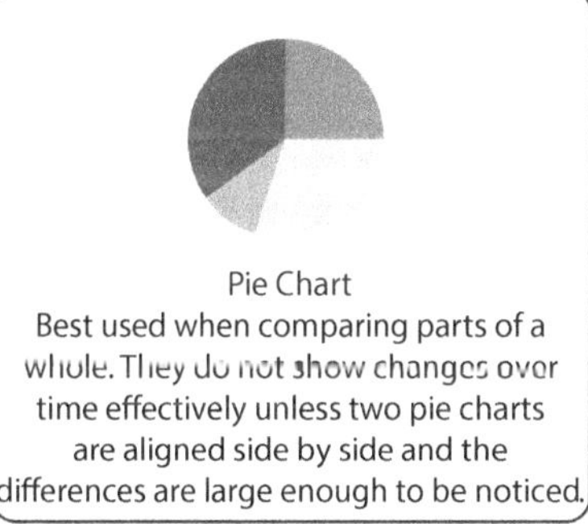

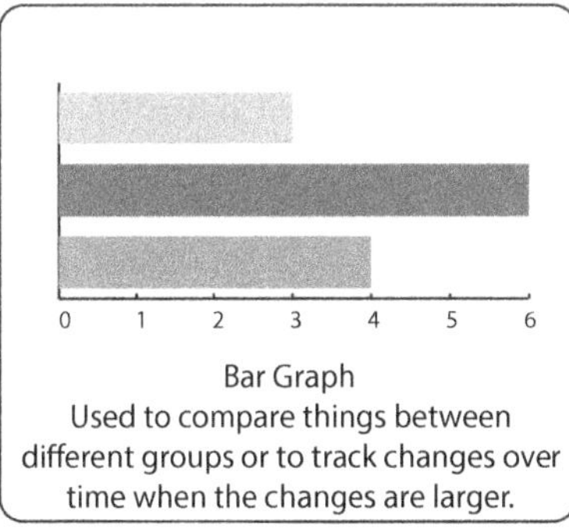

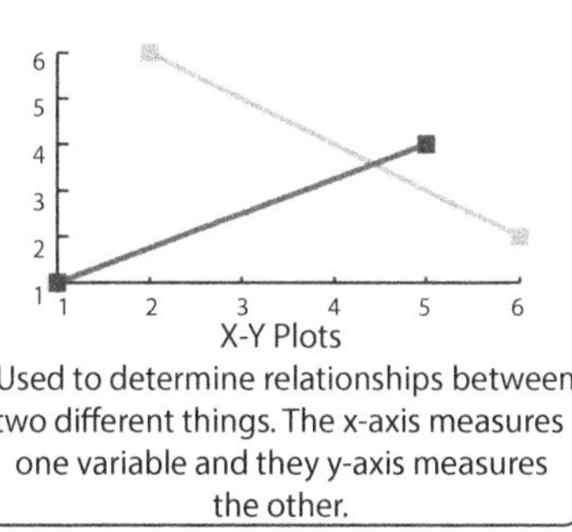

FIGURE 2.4 *Selecting the correct chart or graph*

DOI: 10.1057/9781137523792.0007

Even with the move to more infographics and multimedia presentations, analysts will still need to use standard graphic elements, such as pie charts and bar charts. Katherine Pherson and Randolph Pherson devote an entire chapter of their book, *Critical Thinking for Strategic Intelligence*, to a discussion of how graphics can support analysis.[21] They note that various analytic disciplines, such as military analysis or political analysis, routinely use the same types of graphics to represent data. Figure 2.4, taken from the US Department of Education National Center for Education Statistics,[22] is a simple reference to consult when trying to decide which type of chart or graph to use.

Notes

1 Presentational Tradecraft is a term first used by Geoff Fowler, former Director and Managing Editor of CIA's *World Intelligence Review* (*WIRe*), to describe what analysts should consider when writing for the web.

2 www.poynter.org/extra/Eyetrack/previous.html (Accessed on April 29, 2014).

3 Chip Scanlan, *The Web and the Future of Writing*, Poynter Institute website (Accessed on March 2, 2011).

4 www.useit.com/alertbox/print-vs-online-content.html (Accessed July 30, 2012).

5 Chip Scanlan, *The Web and the Future of Writing*, Poynter.com (Accessed on March 2, 2011).

6 Guillermo Franco, *What Is the Future of Text Online*? Poynter.com (March 16, 2007).

7 Jacqueline Marino, *How Tablets Are Changing the Way Writers Work*, Poynter.com (Accessed on October 2, 2013).

8 Rubén Arcos, "Producing and Consuming Intelligence Products in the Digital Era: The Need for Multimedia Communication," paper presentation.

9 The phrase, "the quick read, the substantial read, and the encyclopedic read," was coined by Mario Garcia, Poynter Institute. Referenced in Chip Scanlan's *The Web and the Future of Writing*, Poynter Institute website (Accessed on March 2, 2011).

10 Nielsen, Jakob as quoted in Marren, Joe. Writing for the Web, Poynter Institute website (Accessed on March 2, 2011).

11 Dube, Jonathan, *Writing News Online*, Poynter Institute website (Accessed on March 2, 2011).

12 This concept in discussed in more detail in Chapter 4, "Establishing a New Paradigm of Collaboration." See the section entitled, "Providing Context."

13 Heath, Chip and Dan, *Made to Stick* (Random House, 2007).

DOI: 10.1057/9781137523792.0007

14 Poynter Institute, EyeTrack2004, multimediarecall, http://www.poynter.org/extra/eyetrack2004/multimediarecall.htm (Accessed on April 29, 2014).
15 See the discussion of personal risk perception factors in Chapter 3, "Communicating Risk Properly" for a fuller treatment of this topic.
16 http://www.nytimes.com/newsgraphics/2013/10/27/south-china-sea/.
17 http://www.nngroup.com/articles/accordions-complex-content/, Hoa Loranger (Accessed on May 18, 2014).
18 http://www.theguardian.com/world/ng-interactive/2014/apr/bangladesh-shirt-on-your-back (Accessed on April 16, 2014).
19 Robin Williams, *The Non-Designer's Design Book* (Berkeley, CA: Peachpit Press, 2008) 3rd edition.
20 See Jennifer Krynin, "Color Symbolism Chart by Culture: Understand the Meaning of Color in Various Cultures Around the World," *About.com Guide*, http://webdesign.about.com/od/color/a/bl_colorculture.htm.
21 See chapter 18, "How Can Graphics Support My Analysis?" in Katherine Hibbs Pherson and Randolph H. Pherson, *Critical Thinking for Strategic Intelligence* (Washington DC: CQ Press/SAGE Publications, 2013).
22 https://nces.ed.gov//whentouse.asp. Graphing Tutorial.

DOI: 10.1057/9781137523792.0007

3
Communicating Risk

John Pyrik

Abstract: *The Intelligence Community (IC) in both the United States and Canada have been trying to heed the demands of intelligence consumers with a product that provides full, true, and plain disclosure but may lack the crucial ingredient of "affect" to effectively communicate risk. This chapter proposes some simple measures in the field of risk analysis which could compensate for the consumer's innate heuristics and risk perception factors. The goal is an analytic product with more impact because it engages both experiential and analytic thinking in dealing with risk perception and communicating risk.*

Keywords: communication; intelligence; risk analysis; risk perception

Arcos, Rubén and Randolph H. Pherson, eds. *Intelligence Communication in the Digital Era: Transforming Security, Defence and Business.* Basingstoke: Palgrave Macmillan, 2015. DOI: 10.1057/9781137523792.0008.

 DOI: 10.1057/9781137523792.0008

Introduction

Intelligence analysts inform decision makers about potential risks to national security and national interests. This is done verbally in briefings and through a variety of written products, one of the best known being the President's Daily Brief (PDB) that is used to brief the President of the United States.

Risk, even within this context, is a broad topic. The Society for Risk Analysis defines risk as "the potential for realization of unwanted, adverse consequences to human life, health, property, or the environment." Consequently, intelligence analysts write papers on a range of issues such as cybersecurity, political instability, and terrorism.

When discussing risk, up until the fall of the Soviet Union, the dominant approach for intelligence analysts was always to describe the *threat from* an actor. Typically, these threat assessments spoke of the *intentions and capabilities* of an adversary like the former Soviet intelligence service, the KGB. The basic formula was intentions plus capabilities equals threat.

$$\text{Intentions} + \text{Capabilities} = \text{Probability of Threat}$$

$$I + C = P\ (\text{threat})$$

The Intelligence Community (IC) began to change this approach after the Cold War ended. While various adversaries continued to pose threats, many more players now were on the field. Their status – friendly, hostile, or neutral – was uncertain and dynamic. As a result, analysts began to talk more about the *threat to* a person, place, or thing. This necessitated thinking about *vulnerabilities* to estimate the probability of a successful attack. Many analysts broadened their concept of risk to include the probability of a successful attack and the impact of the attack, resulting in a new formula.

$$P(\text{threat}) \times P(\text{success}) \times \text{Impact} = \text{Risk}$$

Communicating each component of this equation in an intelligence assessment is a challenge. This challenge is made even more difficult when vague terms such as "likely" and "possible" are employed. Accordingly, efforts have been made over the years to standardize terms and formalize methodology for using probabilistic language.[1]

One aspect, however, has been somewhat neglected – all decision makers don't perceive risk the same way. People view risk differently

DOI: 10.1057/9781137523792.0008

because they process risk information based on their existing beliefs and values.[2]

As writers, we assume that our readers share the same perception of risk as we do, but there may be a "Perception Gap." What we thought was a "low" risk, might be perceived by others as a "high" risk. Perhaps the decision maker had access to additional information which mitigated the risk (or elevated it), but the gap could also be due to personality factors. Analysts begin to address this issue when they ask, "How will my client react?"

Now that the technology exists to create customized products for individual clients, analysts have the opportunity to refine their arguments and present their findings in ways that will be easier for the client to assimilate. The goal is laudable, but why stop there? If possible, should not analysts also adjust their products to diminish the risk perception gap as well?

This chapter examines current practices, their shortcomings, the consequences, and concludes with some ideas that may push ethical boundaries for some analysts.

Communicating risk

In 2005, the Commission on the Intelligence Capabilities of the United States Regarding Weapons of Mass Destruction wrote in its final report that "analysts have a difficult time stating their assumptions up front, explicitly explaining their logic, and, in the end, identifying unambiguously for policymakers what they do not know."

Since that time, intelligence products have gradually been reformatted to highlight the evidentiary basis and list the key assumptions that underpin the analysis. This adds transparency to the analysis, but research suggests that transparency is not enough. As Slovic, Finucane, and Peters have noted:

> We cannot assume that an intelligent person can understand the meaning of and properly act upon even the simplest of numbers ... not to mention ... statistics pertaining to risk, unless these numbers are infused with ***affect***.[3]

The word "affect" in this context means the same thing as emotion. Affect-laden images and language change perceptions of risk. Examples of affect include pictures of starving children and atomic explosions – both highly evocative images.

DOI: 10.1057/9781137523792.0008

Not surprisingly, affect also has an impact on decision making. The **Affect Heuristic** is the tendency to rely on positive or negative feelings evoked by a stimulus to make a decision. A heuristic is a simple procedure that helps find an "adequate, though often imperfect" answer to a difficult question.[4]

One example of the Affect Heuristic is the Halo Affect. In the 1920s, psychologist Edward Thorndike noticed an anomaly in the way commanding officers in the military evaluated subordinate soldiers. Those who were rated high in "physique" tended to be rated high in intelligence, loyalty, and dependability. The correlations were too high, suggesting an unconscious bias.[5]

The Affect Heuristic can be amplified by the Vividness Heuristic and the Availability Heuristic. The vividness of a horrific act like 9/11 not only makes it difficult to forget but tends to inflate our sense of how often terrorist incidents take place. Shark attacks make the news more often than parts falling from an airplane. Because such incidents are more "available," we assume they are more frequent.[6] In both cases, vividness and availability are proxies which our brain uses when it needs a quick method of estimating probability.

The presence of these heuristics is evidence of how easily the human brain can be overtaxed. These short-cuts permit faster decision making and serve as substitutes when rational thinking is too difficult. The key point is that rather than systematically assessing risk factors and mitigating factors, people use "*experiential thinking*" to approximate risk.

The intuitive method is automatic, natural, non-verbal, and usually a reflection of past experiences.

- Memories consist of concrete images, metaphors, and narratives.
- Events evoke positive or negative feelings.
- Feelings guide behavior (decisions "feel right").
- Reactions are instinctual and immediate.
- The cognitive process is rapid and effortless.

The analytic method is conscious, deliberative, verbal, and rational.

- Memories consist of concepts (symbols, words, and numbers).
- Events evoke logical associations (not emotions).
- Logic and reasoning guide behavior. Decisions are based on analysis.
- Reactions are conscious and deliberate.
- The cognitive process is slow and effortful.

FIGURE 3.1 *Two systems of thinking*[7]

DOI: 10.1057/9781137523792.0008

One brain: two systems – intuitive versus analytic thinking

People apprehend reality in two fundamentally different ways, intuitive and analytical. The intuitive method is automatic, natural, non-verbal, narrative, and experiential. The analytic method is deliberative, verbal, and rational.

Each method has its limitations. The experiential system can only guide us in familiar territory. It is also prone to inherent biases and focused on the here and now, rather than the future. Moreover, it is easily manipulated by those who wish to control our behaviors (advertisers, politicians, religious leaders, etc.). The analytic system is slow and, without emotional engagement, a person could lack the incentive or confidence to make a decision or take action.[8]

In essence, there is a high road and a low road to cognition.[10] During cognition, emotion and reason are simultaneously engaged. Emotion (affect) is always present, but as shown in the diagram, the influence of reason can vary from zero (on the left) to a high value (on the right). As Figure 3.2 illustrates:

- Reason takes time and effort.
- Emotion never disappears (entirely).

Communicating with emotion and reason

Ideally, to communicate effectively, one would craft a message to engage both the experiential and analytic systems. This is particularly important in situations where statistics would numb the brain.

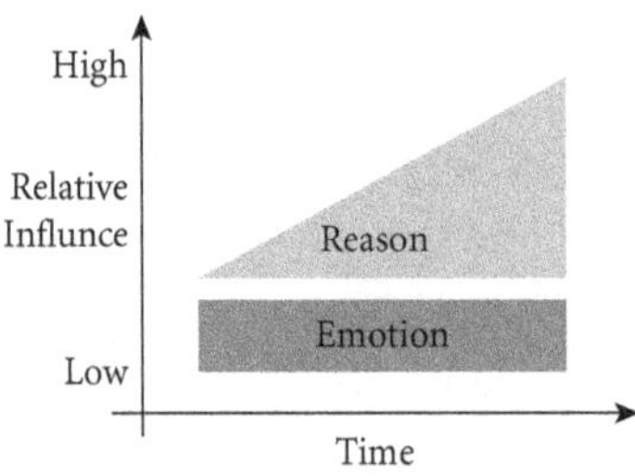

FIGURE 3.2 *Emotion versus reason continuum*[9]

DOI: 10.1057/9781137523792.0008

Soviet dictator Joseph Stalin allegedly once said that "The death of one man is a tragedy; the death of millions is a statistic."[11] So, how can we add tears to statistics?

Examples abound where people have made the abstract more tangible.

- After September 11, many newspapers published biographical sketches of the victims, a dozen or so each day until all had been featured.
- The *Diary of Anne Frank* puts a human face on a tragedy of unimaginable scale.
- Organizers of a rally protesting 38,000 deaths a year from handguns piled 38,000 pairs of shoes in a mound on Capitol Hill.[12]

People have difficulty relating to a number but seeing a pile of empty shoes naturally provokes one to think of the missing owners and an empty pair of kid's shoes seems particularly tragic. For some of the legislators walking up the steps of the Capital Building, the pile of shoes they saw probably made the fact of 38,000 deaths per year more real. The emotional impact may have influenced the way they voted.

Legislators aren't the only ones whose decision making is affected by emotion. In his book, *Thinking Fast and Slow*, Nobel Prize winning psychologist Daniel Kahneman relates the story of an executive who invested millions of dollars in the stock of Ford Motor Company. His decision was based, not on performance statistics, but the fact he had recently attended an automobile show. "Boy, do they know how to make a car!"[13]

The problem for the IC is that a typical analytic product is geared almost entirely to the analytic side of the brain. Writers prefer a rational presentation of the facts. They see an appeal to emotion as evidence of a weak argument. In consequence, by ignoring the emotional dimension, they communicate using only one channel. In short, while research on decision-making has shown that emotional or affective-cognitive factors can influence the process of judgement in many ways, this has not been widely acknowledged or incorporated into the way intelligence is communicated to decision makers.

If today's analytic products are dry as dust and like a movie with no background music, what can be done? Perhaps the ideas and best practices of other professions offer some insights.

DOI: 10.1057/9781137523792.0008

Crude tactics

Warning labels on dangerous products are designed to evoke an emotional reaction. The text and accompanying images are meant to be scary. They must command attention, stimulate memory, suggest consequences, and encourage safe behavior.[14]

This is why manufacturers were forced to add graphic warnings on cigarettes for sale in the United States. The nine images selected by the US FDA include a mottled, brown, diseased lung; an emaciated, hollow-eyed cancer patient; a close-up of a diseased mouth with an open sore and rotting teeth. It was the biggest change in tobacco warnings in 25 years.

While such blatant tactics are effective at triggering emotions, they would be out of place within the staid and conservative IC. Yet, such conservatism must be balanced against the cost of poor risk communication.

Would a more aggressive and "affect-imbued" presentation of risk have changed the outcomes of high-profile failures such as Hurricane Katrina, the MMR vaccine, and Mad Cow disease? In 2012, an earthquake killed 309 people in the town of L'Aquila, Italy. One researcher claimed that when he tried to warn residents prior to the earthquake, he was muzzled by local officials and placed under investigation for causing alarm.[15]

In each of these cases, risk was communicated but misperceived or ignored. Clearly, there is a difficult balance to strike. It is frustrating for analysts that, like Cassandra, they seem cursed to see the future, but are unable to convince others of their vision or rouse them to act.

Better practices

Risk communicators are drawing on studies of behavioral decision-making to understand risk perception and how people make choices about risk.[16] However, risk communication is still a complex process. Many factors need to be taken into account for effective risk communication to occur, including good science, economic, social, cultural, ethical, political, and legal considerations.[17] Preliminary findings suggest that perceptions of risk are issue dependent, and limited progress has been made in producing more effective risk communication programs that meet the needs of both the risk communicator and the recipient.[18]

DOI: 10.1057/9781137523792.0008

In communicating risk, however, a general rule of thumb is to begin with a qualitative term, then follow-up with a quantitative term.[19] If the reader agrees that there is a "remote" possibility, the argument is half-way won. The writer then need only ensure that the term "remote" is mutually understood to mean the same thing. This can be achieved through the use of quantitative terms, for example, by adding the sentence: "By remote, we mean a probability of less than 5%."[20]

According to a systematic review of medical patients, they have a more accurate perception of risk if qualitative terms are paired with quantitative terms. In other words, describing the chances of a bad event as "rare" is acceptable as long as the term is then linked to a frequency like "1 in 1000."[21]

In addition to a term of probability (expressed qualitatively and quantitatively), a statement about risk should describe the *type of risk* and provide *supporting evidence*. "The possibility is remote, but Denmark could send troops to occupy Hans Island, a territory claimed by Canada, given recent public statements by the Danish Prime Minister."

To help the reader understand, the writer may expand on both the type of risk and describe the supporting evidence in greater detail. The writer might, for example, address such questions as:

- How good is the data?
- Is there a possibility of deception?
- How good is the model?
- Are there other interpretations/explanations?
- Is there consensus among experts on the leading hypothesis?

Communication could be further improved if the numbers were infused with ***affect*** as it engages ***experiential thinking***. A subtle method involves shifting from terms that are abstract to those that are more concrete, i.e. relatable. If 10% of patients die from a certain drug, it may seem safe enough. However, if one in ten patients dies, many perceive the risk as higher. Expressing probability in terms of frequency (10 out of 100) rather than as a percentage (10%) apparently affects our estimation of risk, possibly because the former is more concrete than the latter.[22]

Another effective illustration of the use of affect was provided by Richard Peto in 1980 who delivered this warning about the dangers of smoking.

> Among an average 1000 young men who smoke cigarettes regularly – about 1 will be murdered, about 6 will be killed on the roads, and about 250 will be killed before their time by tobacco.[23]

DOI: 10.1057/9781137523792.0008

The number "1000" is large but manageable; most people can picture it (e.g. a small hockey arena filled with people). The term "young men" might have particular resonance with the target audience of this message. The word "murder" and the image of being "killed on the road" are vivid. Ultimately, the reader is left with a powerful statistic: one murdered, six killed in car accidents, but 250 killed by smoking! Someone who felt driving was dangerous, now realizes that smoking is more than *80 times* more dangerous.

A common tactic used by journalists involves "humanizing" a story. A deadpan report that wildfires have destroyed 1,000 homes elicits one reaction. Interviewing a victim who barely got out with their life and their cat, puts a human face to the story and elicits an empathetic reaction.

Pairing qualitative and quantitative terms, going from abstract to concrete, and adding affect are all good strategies. A further refinement would be to consider how the specific intended audience is likely to perceive risk and craft the message accordingly.

Best practices

Risk perception factors make our fears go up or down. They combine to create an overall perception of a threat and are inputs into the overall system of Affective Risk Perception.[24]

Four of the most important Risk Perception Factors (Trust, Benefit, Choice, and Control) are described in Figure 3.3. The influence of these factors on our perception of risk varies. Life experience, education, lifestyle, and culture can amplify the effect or even reverse it. If you have never driven a motorbike, for example, you would probably be more comfortable as a passenger than as the driver. In this instance, more control would make you more nervous.

The adept intelligence analyst could take these risk perception factors into account when writing for a specific client. If it was known, for example, that the client was likely to under-perceive the risk, the analyst could modify his product accordingly. Each of these four factors can be "tweaked" by an analyst to change a reader's perception of risk.

Trust is the first factor listed because it is probably the most influential. Earn someone's trust and you can probably convince them of almost anything. Fraudsters exploit this through something called "affinity

DOI: 10.1057/9781137523792.0008

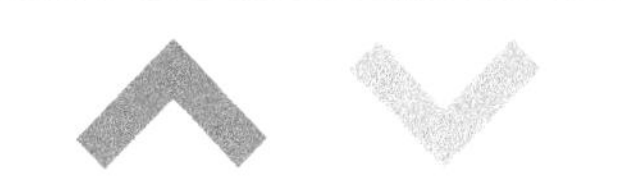

FIGURE 3.3 *Four risk perception factors*

fraud." While most people are leery when a stranger propositions them with a get-rich-quick scheme, they let their guard down when it is someone they trust. It need not even be deep trust; it can simply be the fact that they share something in common with the fraudster (e.g. a common ethnicity, religion, or profession). "Trust me!" says the fraudster. "I'm just like you." This affinity allows the fraudster to circumvent a person's normal skepticism.

The power of this factor is clear when you think about how hard it is to regain a person's trust once you lose it. This is a lesson many companies and governments have learned the hard way. The CIA lost credibility when it mistakenly thought Iraq still had weapons of mass destruction (WMD). One of the results was the Intelligence Reform Act and Terrorism Prevention Act of 2004.

If a client has "low trust" in your analytic products, they are not inclined to believe what you have to say. No matter the revelation, the client unconsciously looks for reasons to disbelieve your findings. If this

DOI: 10.1057/9781137523792.0008

were the case, it would be reasonable for the analyst to attempt to be more persuasive.

Traditionally, analysts have responded to situations of "low trust" with reports and memos that have more data, more rigour, and more on the negative consequences of ignoring the reported findings. To improve rigour, for example, analysts are now using Structured Analytic Techniques (SATs).[25]

A key tactic, post 9/11, was more transparency in the analytic process.

> Tell me what you know. Tell me what you don't know. And then, based on what you really know and what you really don't know, tell me what you think is most likely to happen.
>
> – *Colin Powell, Opening Remarks before the Senate Governmental Affairs Committee, Washington, DC, September 13, 2004.*

Benefit is the second risk perception factor listed. To recap, the higher the perceived benefit, the lower the perceived risk.

Examples of the concept abound, especially in relation to investment decisions. Ask someone to invest in a shady deal that will return 10% and they will likely balk. Raise the rate of return to 100% and they hesitate. Make it 1000%, and they will rationalize that "it is worth the gamble." Nothing has changed but the expected benefit, yet this has the effect of lessening the perception of risk.

Imagine a situation where a policymaker is personally invested in seeing a certain outcome. Regrettably, the analysis shows that this outcome is increasingly unlikely. The analyst presenting these findings is fighting an uphill battle.

This is not an uncommon situation. In 1968, President Lyndon Johnson famously complained that "intelligence guys" were complicating his life.

> Let me tell you about these intelligence guys. When I was growing up in Texas we had a cow named Bessy. I'd go out early and milk her. One day, I'd worked hard and gotten a full pail of milk but I wasn't paying attention and old Bessy swung her shit-smeared tail through the bucket of milk. Now you know, that's what these intelligence guys do. You work hard and get a good program or policy going and they swing a shit-smeared tail through it.[26]

Johnson was suggesting that when an analyst is delivering bad news, there is a lack of benefit for the client. This might increase their perception of risk.

Conversely, if the situation had "high benefit" for the client, it might lower his or her perception of risk. Recently, for example, Western governments have been reported to pay for the release of hostages held by terrorist groups. The risk of encouraging further kidnappings seems to have been overridden by short-term benefit. If an analyst knew the tendency of decision makers, he might point out second-order and long-term complications that reduce the overall benefit. This could have the effect of restoring the perception of risk to a more realistic level.[27]

Choice is the third risk perception factor. If someone feels his choice is forced, they consequently perceive more risk. The decision to evacuate diplomatic staff from an embassy in a city with civil unrest could be the right thing to do, but it could also be a costly over-reaction. If an analyst presented this scenario as "Evacuate, yes or no?" the decision maker might freeze. To lower their risk perception, the analyst could suggest an intermediate course of action that buys more time before a tougher choice must be made.

Control is the fourth and final risk perception factor. If the client had "high control" over a situation, the concern would be that he would underestimate the risk. To compensate, the analyst could stress the complexity of the situation, the constraints on action, and the forces which would resist or counteract any actions the decision maker might take.

In summary, the more that is known about the decision maker, the better the message can be customized to his or her unique risk perception factors.

The ethical conundrum

Purists, schooled to write in terms that were dry as toast, would regard a message crafted to appeal to emotion as manipulative and likely unethical. They would say that such messages target unconscious processes and diminish a person's ability to make a free and rational choice.[28]

Examples abound of using fear to manipulate consumers, sometimes creating unnecessary anxiety.[29] Efforts are criticized even when the intention is good.[30] Some, for example, argue that government should only present "factual" information on risks to the public.[31]

Yet all messages are manipulative to some extent. A lawyer reviewing a link chart depicting a complex set of transactions didn't like the fact that the icon of the person at the centre of the chart was twice the size of the other icons on the page. "Make his head the same size as everyone else – otherwise it looks like we're trying to tell the police who to investigate."[32]

DOI: 10.1057/9781137523792.0008

In the private sector, "brands" are a means of manipulating buyers. In a similar sense, the reputations of government departments help them "sell" their products and services.

The suggestion here is not to "play with the numbers" or "sex up the file" but merely to gently counterbalance the natural tendencies of the client. Theoretically, such a customized message would provide the client with a more accurate sense of the risk.

A final consideration is timing. For intelligence to be welcomed and have an impact, it must arrive at the right time.[33] If one is too early, it is an uphill battle to get the attention of a decision maker. If one is too late, the decision maker has already moved on to other problems. The golden window is when decision makers first become seized with the problem, but before they have made up their minds. Analysts need to know what stage the decision maker is in, then adjust their communication strategy accordingly.

Notes

1 One of the earliest efforts to standardize terms of uncertainty and probability was made in 1964 when Sherman Kent wrote an article entitled "Words of Estimative Probability" for the CIA's internal publication, *Studies in Intelligence*.

2 Multiple sources:
Slovic, P., Finucane, M. L., Peters, E. and MacGregor, D.G. "Risk as Analysis and Risk as Feelings: Some Thoughts about Affect, Reason, Risk, and Rationality". *Risk Analysis* 24 (2004):311–322. doi: 10.1111/j.0272-4332.2004.00433.x
Slovic, P. "Trust, Emotion, Sex, Politics, and Science: Surveying the Risk-Assessment Battlefield". *Risk Analysis* 19 (1999):689–701. doi: 10.1111/j.1539-6924.1999.tb00439.x
Fischhoff, B. "Risk Perception and Communication Unplugged: Twenty Years of Process". *Risk Analysis* 15 (1995):137–145. doi: 10.1111/j.1539-6924.1995.tb00308.x
Morgan, M. G. et al., "Communicating Risk to the Public". *Eviron. Sci. Technol.* 26 (1992): 2048–2056. doi: 10.1021/es00035a606
Slovic, P. "Perception of Risk". *Science* 236 (1987): 280–285. doi: 10.1126/science.3563507.

3 Slovic, P.; Finucane, M. L.; Peters, E. and MacGregor, D. G. "Risk as Analysis and Risk as Feelings: Some Thoughts about Affect, Reason, Risk, and Rationality". *Risk Analysis* 24 (2004): 311–322.

4 Daniel Kahneman, *Thinking, Fast and Slow* (New York: Macmillan, 2011), 98.

5 K. Rasmussen, "Halo Effect" in N. J. Salkind & K. Rasmussen (eds.) *Encyclopedia of Educational Psychology*, 1 (Thousand Oaks, CA: Sage Publications, 2008).

6 Lichtenstein, S.; Slovic, P.; Fischhoff, B.; Layman, M. & Combs, B. "Judged frequency of lethal events". *Journal of Experimental Psychology: Human Learning and Memory* 4, (1978): 551–578.

7 Slovic, Paul; Finucane, Melissa L.; Peters, Ellen and MacGregor, Donald G. "Risk as Analysis and Risk as Feelings: Some Thoughts about Affect, Reason, Risk, and Rationality". *Risk Analysis* 24 (2), (April 2004): 311–322,. Article first published online: April 13, 2004. doi: 10.1111/j.0272-4332.2004.00433.x.

8 Slovic, *Risk as Analysis*, 311–322.

9 Source of data: Ross Buck and Whitney A. Davis, "Marketing risk: Emotional appeals can promote the mindless acceptance of risk," in Sabine Roeser (ed) *Emotions and Risky Technologies* (Berlin: Springer, 2010): pp. 61–80.

10 J. E. LeDoux, *The Emotional Brain: The Mysterious Underpinnings of Emotional Life* (New York: Simon & Schuster, 1996), 161–164.

11 Fred R. Shapiro, *The Yale Book of Quotations* (New Haven: Yale University Press, 2006), 724.

12 "38,000 shoes stand for loss in lethal year," *The Register-Guard*, September 21, 1994, 6A.

13 Kahneman, *Fast and Slow*, 12.

14 Colin Poitras, "Graphic Cigarette Warnings Evoke Important Emotions," *UCONN Today* (February 2, 2011). Quoting Ross Buck.

15 Nick Squires, "Italian Earthquake: Expert's Warnings Were Dismissed as Scaremongering," *Telegraph* (April 6, 2009).

16 Maibach E. & Holtgrave, D. R. "Advances In Public-Health Communication". *Annual Review of Public Health* 16 (1995), 219–238.

17 *Commission on Risk Assessment and Risk Management,* Framework for Environmental Health Risk Management, *by* Gilbert Omenn et al. *(Washington, D.C., United States Government Printing Office, 1997), 64.*

18 Faulkner, Hazel & Ball, David. "*Environmental Hazards and Risk Communication*". *Environmental Hazards 7 (2007):71–78.*

19 Peter Wiedemann, Martin Clauberg, and Franziska Börner, *Risk Communication for Companies* (2010): 34.

20 For a fuller discussion of the use of probabilistic language in intelligence analysis see Katherine Hibbs Pherson and Randolph H. Pherson, *Critical Thinking for Strategic Intelligence*, chapter 17: "How Should I Portray Probability and Levels of Confidence" (Washington DC: CQ Press/SAGE Publications), 2013.

21 L. J. Trevena, H. M. Davey, A. Barratt, P. Bulow, & P. Caldwell, "A systematic review on communicating with patients about evidence," in *J Eval Clin Pract* 1 (2006):13–23.

DOI: 10.1057/9781137523792.0008

22 Slovic, Paul; Monahan, John & MacGregor, Donald. "Violence risk assessment and risk communication: The effects of using actual cases, providing instructions, and employing probability vs. frequency formats". *Law and Human Behavior* 24(3), (2000): 271–296.

23 World Health Organization, *The Health of Young People* (Geneva, 1993). Quoting Richard Peto.

24 David Ropeik, *How Risky is it, really?* (New York: McGraw-Hill, 2010), 69.

25 The best description of Structured Analytic Techniques is provided in Richards J. Heuer Jr. and Randolph H. Pherson, *Structured Analytic Techniques for Intelligence Analysis*, 2nd Edition (Washington DC: CQ Press/SAGE Publications, 2015).

26 Jervis, Robert. "Why intelligence and policymakers clash". *Political Science Quarterly* 125(2), (2010): 185.

27 An investigation by *The New York Times* found that al Qaeda and its direct affiliates have taken in at least $125 million in revenue from kidnappings since 2008, of which $66 million was paid just last year. Callimachi, R. "Paying Ransoms, Europe Bankrolls Qaeda Terror", *New York Times*, July 29, 2014.

28 Tom Beauchamp, "Manipulative Advertising" in Tom Beauchamp and Norman Bowie editors, *Ethical Theory and Business*, 3rd Edition (Englewood Cliffs, NJ: Prentice Hall, 1988), 694.

29 Benet, Suzeanne; Pitts, Robert E. and LaTour, Michael. "The appropriateness of fear appeal use for health care marketing to the elderly: Is it OK to scare granny?" *Journal of Business Ethics* 12(1), (1993): 45–55.

30 Arthur, Daimien and Quester, Pascale. "The ethicality of using fear for social advertising". *Australasian Marketing Journal* (AMJ) 11(1), (2003): 12–27.

31 Ross Buck and Whitney A. Davis, "Marketing risk: Emotional appeals can promote the mindless acceptance of risk," in Sabine Roeser (ed) *Emotions and Risky Technologies*, (Berlin: Springer, 2010): pp. 61–80.

32 Author's personal experience while working as an analyst at Canada's anti-money laundering agency in 2004.

33 Jervis, *Clash*, 196.

DOI: 10.1057/9781137523792.0008

4
Establishing a New Paradigm of Collaboration

Randolph H. Pherson

Abstract: *New digital technologies are opening the door to establishing of a new paradigm of collaboration for producing and delivering intelligence analysis to decision makers. With the use of web-based wikis, analysts can draft articles jointly, blurring the lines between all-source analysts, collectors, and even decision makers. Access to drafts and finished products can be defined at several levels by varying who is given the authority to "read" or "write" in the document. Key supplemental information can be presented through pop-ups, side bars, or hyperlinks. This would allow the reader to focus on what is most important, including the use of structured analytic techniques. Avatar-based collaboration platforms would also facilitate greater interaction and collaboration, allowing analysts to coordinate and brief their papers without ever leaving their desks.*

Keywords: analysis; avatars; collaboration; intelligence; structured analytic techniques

Arcos, Rubén and Randolph H. Pherson, eds. *Intelligence Communication in the Digital Era: Transforming Security, Defence and Business.* Basingstoke: Palgrave Macmillan, 2015. DOI: 10.1057/9781137523792.0009.

The vision

The development of powerful new digital technologies for communicating information and analysis has greatly enhanced both the ability and the prospects to engage in synchronous and asynchronous collaborative activities. The production of an analytic product no longer must be defined as a serial process with a lead analyst crafting an initial draft, incorporating contributions from colleagues, coordinating and revising subsequent drafts, and submitting the paper to editors and managers to review. With the use of a wiki (a website that allows collaborative editing of its content and structure by its users), the drafting process can become an act of co-creation involving a much broader range of participants.[1] In such an environment, even the recipients of the analysis could become contributors. Analytic products would no longer be generated and reviewed within a single agency stovepipe, although a lead author and a lead publishing office or agency probably would still be required for purposes of accountability and efficiency.

With the advent of a "pull" system of distribution involving the posting of analytic products on a website for any authorized reader to access, the potential audience for each product would grow exponentially. Producers would no longer have to estimate who would be interested in that topic and should be put on a distribution list; instead, those who are interested – and appropriately approved for access – would seek out what most interests them. By "publishing" analytic products in digital form as opposed to hard copy, they also can be readily updated as events and consumer demand warrant. This process of periodic refreshment will also make the products more attractive to a wider base of consumers.

Increasing Efficiency. The move to analytic products generated through a collaborative process would have several corollary effects. Within the Intelligence Community, for example, several agencies will often write on the same topic; each author will often cover much of the same ground because he or she needs to establish a baseline for his or her assessment. The article will usually be tailored to reflect the particular contributions that agency can make to the analysis or reflecting what that particular agency has collected on the subject. The policymaker then inherits the task to read each of these overlapping products and synthesize the various inputs to come up with an overall picture of the situation.

A *wiki*-based system would save both the analysts and the consumers of the analysis considerable time by avoiding duplicative drafting and

DOI: 10.1057/9781137523792.0009

building analytic consensus early on in the process. The description of information provided by key sources as well as other baseline information would only have to be drafted once, allowing contributors to focus their attention on where they add the most value to the analysis. The process also ensures that everyone is working from the same sheet of music and shares the same overarching view or *meta*-analysis of the event.

The system would be most suitable to the analytic papers at each end of the product spectrum – either longer range, multi-page Intelligence Assessments and National Intelligence Estimates or quick turnaround Situation Reports, Warning Reports, and Spot Assessments. It would have less value as a platform for generating current intelligence reporting or other short, often one- or two-page reports that often take only a few days to craft. The platform would also be an ideal place to post background material and basic intelligence reporting that would not require frequent updating.

Wikipedia is the most well-known commercial product that attempts to accomplish many of these objectives but its anonymity and the inadequate sourcing of many of its entries severely limit its utility. The Intelligence Community has developed its own, classified version of *Wikipedia* called Intellipedia and it could serve as the necessary platform for implementing the system proposed in this chapter (see Figure 4.1).

Dealing with differences. The greatest benefit, however, could well be that the ability of community would speak with a coordinated, consolidated, and coherent voice on key topics of interest would be significantly enhanced. Analysts would find it easier to work out their difference early on in the drafting process; they could share their evidence and challenge each other's assumptions before analytic positions are firmly established. When major differences of opinion occur, a frequent culprit is the lack

Wiki-based Intelligence Production	*Wikipedia*
Name and affiliation of drafters provided	Anonymous drafters
Comprehensive source citations	Inconsistent sourcing
Sources evaluated and rated	No source evaluation
Focus on issues of current intelligence interest	Encyclopedic
Formal review and editing process	Informal review and editing
Corporate product	Produced by individual(s)

FIGURE 4.1 *How does this* wiki-*based intelligence production compare to* Wikipedia*?*[2]

DOI: 10.1057/9781137523792.0009

of data on that particular issue. When data is lacking, assumptions must be made, and the potential to default to organizational positions or fall prey to cognitive traps grows. In the traditional landscape of stove-piped analysis, the result often is a series of footnotes that defend each agency's line of analysis.

With a *wiki*-generated product, the result is likely to be much different. Differences could be resolved before positions become rigid and the overall drafting process would be much more efficient. If analysts succeed early on in exposing the root causes of their differences, then it becomes fairly easy to craft language that states that most analysts explain the phenomenon in this way because of this evidence but other analysts offer a different explanation because they put greater weight on other evidence or make other assumptions. Most consumers of analysis say that this second approach is more helpful because it focuses attention on the basis for a difference, leaving the door open for one or the other explanation to emerge as more correct as additional reports are received and the analysts' understanding of the issue grows.

Blurring organizational boundaries. *Wiki*-based analysis by its very nature blurs the lines between the role of an all-source analyst, an analyst in a collection agency, a collector, and even a decision maker. The key to producing solid analytic products is to tap the expertise of every knowledgeable person on the topic, not just the all-source analyst. As a former National Intelligence Officer for Latin America who managed the production of many national estimates, the author quickly learned that repositories of true expertise could reside at any agency. How long an analyst had worked an account was usually a much better determinant of expertise than which agency they represented.

A collector or clandestine service reports officer who had worked a particular target for ten years almost certainly had more informed insights on a foreign leader's personality and likely behavior than an all-source analyst recently assigned to that country. Those working in the field and those closer to the target invariably are a valuable – and often ignored – primary source of information and insight. Much the same is true for policymakers. They can tap their first-hand experience in dealing with senior foreign officials as well as decades of exposure to how things work in a particular culture.

Moreover, analysis in the US intelligence community is transitioning from a mental activity done predominantly by a sole analyst to a

DOI: 10.1057/9781137523792.0009

collaborative team activity.[3] The driving forces behind this transition include:

- The growing complexity of international issues and consequent requirement for multidisciplinary input to most analytic products.
- The need to share more information more quickly across organizational boundaries.
- The increased dispersion of expertise, especially as the boundaries between analysts, collectors, and operators become blurred.
- The need to identify and evaluate the validity of alternative models for understanding a problem.

A *wiki*-based system of analytic production makes it possible to gather "the best from the brightest." It is oblivious to which agency the insights come from, seeking only the analysis that best stands up to scrutiny regardless of the source.

The process

The production of *wiki*-based analysis requires that both producers and consumers of the analytic products have access to the same operating system. In the commercial world and many parts of government, this is easily accommodated by establishing access-controlled portals or websites on the Internet. When dealing with classified information, the challenge becomes more daunting. A common practice is to stratify classified products into three categories: (1) Unclassified or For Official Use Only documents that are posted on a protected unclassified network such as the Department of Defense's Nonsecure Internet Protocol (IP) Router Network (NIPRNet), (2) documents classified up to the Secret level that are posted on secure networks such as the DoD's Secret Internet Protocol Router Network (SIPRNet), and (3) documents up to the Top Secret level that are posted on DoD's Joint Worldwide Intelligence Communications System (JWICS). In such classified systems, much more attention must be devoted to who has access to which levels of classification and document control. Unfortunately, such stratification also makes it harder for the customer to access critical information, as he or she has to switch back and forth across systems.

Generating the analysis. Once an appropriate platform has been designated for producing the draft, the next step is to establish a team

DOI: 10.1057/9781137523792.0009

of drafters, nominate a key drafter, and decide who will be contributing additional information or analysis to the draft. The names of all the participants along with their affiliations and contact data should be provided at the bottom of the document or in a pop up box. The next and most critical step is for the team members to establish some "rules of the road" for their activity. Such rules would include:

- Deadlines for submitting (or entering) contributions.
- Deadlines for generating a completed draft.
- A process for capturing all source material.
- A strategy for buttressing the draft with graphics or other supporting media.
- Agreement on which office or agency will have responsibility for putting the paper through the final review and editing process.

Two other issues most effectively addressed at this stage of production are whether the team needs to reach out to others for expertise and whether structured analytic techniques should be used to help generate the analysis.[4] A simple rule that most readers greatly appreciate is to include a footnote at the bottom of the article that lists the names of the drafting team and their affiliations and then states: "In preparing this analysis, the drafting team reached out to X and used the following structured techniques: Y and Z."

Often some of the key information needed to draft a paper does not reside in the analyst's computer nor could it be found easily by doing a Google or key word search. The information may reside at a field office or in another organization or university, but it has not been reported or published and is not available electronically. In such cases, it is smart to ask oneself "Who is most likely to possess the information I need, and what is the most efficient way to access that information?"

Most analytic papers will benefit from the application of at least one Structured Analytic Technique.[5] Some of the most commonly used techniques are:

- *Structured brainstorming:* a brainstorming process using sticky notes for generating new ideas and concepts often used to kick off a multiple-scenarios exercise.
- *Key assumptions check:* a systematic effort to make explicit and question the assumptions that guide an analyst's interpretation of evidence and reasoning about any particular problem.

DOI: 10.1057/9781137523792.0009

- *Analysis of competing hypotheses (ACH):* the identification of a complete set of alternative hypotheses, the systematic evaluation of each, and the selection of the best-fitting hypothesis (or hypotheses) by focusing on information that tends to disconfirm each hypothesis.
- *Indicators:* a pre-established set of observable phenomena that are periodically reviewed to help track events, spot emerging trends, and warn of unanticipated changes.

Structured techniques enable collaboration and are perfectly suited to *wiki*-based analytic production. They can guide the dialogue between analysts with common interests as they share evidence and alternative perspectives on the meaning and significance of this evidence. Just as these techniques provide structure to individual thought processes, they can also organize the interaction of analysts within a small team. Because the thought process in these techniques is transparent, each step in the technique prompts relevant discussion within the team. Discussion under these circumstances can generate and evaluate even more divergent information and new ideas than a group that does not follow a structured process.

Once a draft of the paper has been completed, the next step is to subject the draft to peer review and a self-edit. The lead author or any team member can be designated to perform the self-edit using the Critical Thinkers Checklist or other more detailed checklists described in chapter 9, "How Do I Know that I Am Finished?" in Pherson and Pherson's *Critical Thinking for Strategic Intelligence*.[6] The drafting team should also coordinate the paper with those who did not participate in the drafting process but have a stake in what was said, or for more formal publications, are required to coordinate on a draft. This can be accomplished efficiently in the *wiki* environment by employing the Track Changes editing software.

The analysis now is ready for final review and pre-publication editing. The question of which office or agency should conduct that final review and pre-publication edit should have been addressed in the initial "rules of the road" session. In a *wiki* environment, this step is basically a quality control exercise because the analysis will be posted on a communal website and not as a hard copy document with an agency logo or agency seal on the cover. The document should display in a readily accessible location all members of the production team and their affiliations: the

DOI: 10.1057/9781137523792.0009

lead author, contributors, reviewers, and editors. It should also state when the article was drafted and last updated.

Posting the analysis. Several strong arguments can be made for posting the finished intelligence article on the same website as that which was used to generate the paper:

- **Efficiency:** The article can be "published" and distributed to all appropriate readers with only a click or two of the mouse. Because the website relies on a "pull" and not a "push" architecture, the need to create unique distribution lists for each product disappears.
- **Security:** Only those who are authorized to access the website and have appropriate clearances would be allowed to read the article.
- **Accessibility:** One of the unique benefits of using a web environment to prepare an analysis is that interested parties with a "need to know" but who are not members of the drafting team would be able to watch the soup being made as the analysts develop the draft.

The ability to observe "works in progress" can be accomplished by imposing an architecture on the website designating four different levels of access. This is best illustrated by a series of concentric circles (see Figure 4.2).

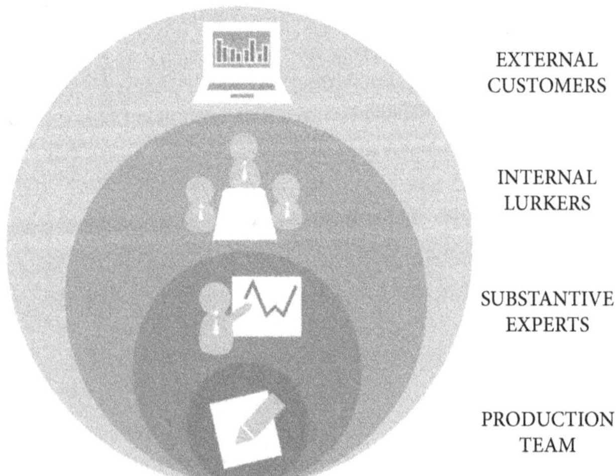

FIGURE 4.2 *Concentric circles of access*
Source: Globalytica, LLC, 2015

DOI: 10.1057/9781137523792.0009

- **Production Team:** The first circle consists of the drafting and production team who have both "read" and "write" access to the file. This team should be defined at the start of the project; including the lead author, contributing authors, and appropriate product reviewers.
- **Substantive experts:** The second circle consists of other analysts, collectors, researchers who might provide inputs to the process (expert comments, additional information, context) but do not have "write" permission. This group also includes colleagues who assist in the use of a structured analytic technique, for example, by participating in a Key Assumptions Check.
- **Internal lurkers:** The third circle consists of those with appropriate clearances within the Intelligence Community (or perhaps the employees of the company in the commercial world) who can observe the soup being made but cannot "write" or input into the process. All managers not involved in product review also fall into this category.
- **External customers:** The fourth and final circle consists of the policymakers and decision makers for whom the analysis is intended. They only have access to a database containing completed papers or finished intelligence products and have no visibility into the making of the stew. As noted before, these papers should include dates for when they were initially published and later updated.

Another major advantage of web-based production is that the analysis can be quickly revised or updated. For example, a Situation Report that summarizes key events or key items of information relating to a crisis has traditionally been published on a set schedule usually once a day or sometimes several times a day. With digital production, the chronology of events can be updated far more often, with less critical, and more dated, data points dropping off the beginning of the chronology as new events are added to the list. Another advantage would be the ability to augment the evidence or include new sourcing for a key assertion or analytic judgment as new information or new reports are acquired.

Providing context. Traditional, hard copy reports are seriously constrained in terms of how much information they can convey by the need to restrict page counts. A busy decision maker who is severely time constrained can only process so many pages in a morning briefing.

DOI: 10.1057/9781137523792.0009

Once the formal business day has begun, he or she usually is even more time restricted, often only capable of processing paper consisting of no more than a page. Sophisticated consumers, however, often would like to have additional information, particularly on topics with which they are personally engaged. For example, they will often want to ask: How good is the information? Who is taking action? What is your level of confidence in the analytic judgment? What key assumptions underlie the analysis?

With *wiki*-based production, these questions can be addressed through the use of pop-ups, side bars, or hyperlinks. The banner across the top of the page, for example, could include some of the buttons listed below. Some of the buttons would be standard such as the Production Team; others might be included only when they are populated with data by the analysts:

- **Production team:** Lists the authors, contributors, reviewers and their affiliations. Might also state who the team reached out to for expertise in preparing the analysis and what structured techniques were used to generate the analysis.
- **Key assumptions:** Lists the results of a Key Assumptions Check exercise showing solid, caveated, and unsubstantiated assumptions that formed the foundation of the initial analysis. Unsupported assumptions would be included on the list because readers might incorrectly be assuming the same thing.
- **Indicators:** If a previous analysis included a list of indicators, signposts, or "things to watch for," clicking on this button could display how many of those indicators have been observed since the initial analysis was published.
- **Timelines and chronologies:** Provides either a timeline or a chronology to guide the reader through a complex analysis.
- **Maps and charts:** Pops out a map when the article mentions several place names to orient the reader. Similarly, charts that are not integral to the story can be provided as ways to display additional background information efficiently.
- **Biographic summaries:** Lists key individuals discussed in the article; the reader could click on the person of interest and read their biographic profile.
- **Critical information gaps:** Lists critical gaps in the reporting and addresses how those gaps might impact on the key judgments.

DOI: 10.1057/9781137523792.0009

- **Collection (or research) requirements:** Provides a list of outstanding collection (or research) requirements that have been levied relating to the topic and what new tasking may have been recently initiated.
- **Source summary statement:** Describes the credibility of the sources used to produce the analysis and the team's overall level of confidence in both the sourcing and the analysis.
- **Glossary of terms:** Provides a list of key words and acronyms used in the document and their definitions and expansions.
- **Related production:** Lists important previously published articles on the topic.
- **Related SMEs:** Lists experts in the field, both internal and external to the community, and their contact data if appropriate.

Few customers would click on all the buttons but most probably would click on at least two if they were directly involved in working that issue. The inclusion of such buttons would also force more analytic rigor into the analytic process. Drafting teams would have to decide whether to conduct a Key Assumptions Check to populate that button or to leave that field blank. Similarly, they would be spurred to consider what impact key information gaps might have had on the analysis or how confident they are in their analytic judgments.

Another advantage of a *wiki*-based production system would be the ability to create libraries of useful databases – or give ready access to such databases that already exist. One example of such a database would be the establishment of an Anomaly Database. When analysts came across a particular item of information that did not seem to fit, imagery that cannot be explained, or new developments that challenged categorization, they could post it to the Anomaly Database to see if any other analysts had encountered – or were puzzled by – the same phenomenon. A chat function could be attached to the database so analysts could exchange views, theories, and concerns about the data.

Given the growing use of Structured Analytic Techniques across the community, analysts would benefit greatly from the creation of a database that captured the results of using the techniques for past problems. Such a database, for example, could contain compilations of lists of:

- Indicators used in previous analytic products.
- Alternative hypotheses generated by various Multiple Hypothesis Generation exercises.

DOI: 10.1057/9781137523792.0009

- Relevant information and alternative hypotheses used in previous Analysis of Competing Hypotheses (ACH) exercises.
- Key assumptions generated on related topics by Key Assumptions Checks.
- Analytic traps and mindsets that have led to previous analyses being wrong.
- Lists of mistakes to watch for that should be incorporated into a Premortem Assessment.

Anticipated impact

Movement toward a new paradigm of collaboration in the production of analytic products will have a major impact on how consumers of analysis view the analytic products they receive. They increasingly demand – and such a transition would greatly facilitate – the ability to provide analytic support on a more timely basis and to deliver products far more efficiently as part of a program of continuous, 24/7 coverage.

Also beneficial from the consumer's perspective would be the move to more collaborative analytic products that reflect the view of all the experts engaged on an issue – including all-source analysts, single source analysts, and even collectors. The blurring of organizational boundaries and the move away from stove-piped analytic products would significantly enhance the ability of a policymaker or decision maker to process and absorb the analysis. Care would have to be taken, however, to avoid the trap of least common denominator analysis. When differences exist, they should be recorded and the reasons for the differences should be explained.

Moving from a "push" to a "pull" system of analytic production could also have serious downsides. In the Intelligence Community, a key lesson learned over the past two decades has been the value of deploying briefers as trusted interlocutors to deliver analytic products and establish a dialogue between the intelligence consumer and the producer. Under no means should this new paradigm of collaborative *wiki*-based analytic production be allowed to undermine these networks of personal relationships and trust-building that provide a human face to the analysis.

One solution to this dilemma would be to maintain the primacy of trusted interlocutors (or designated briefers) but deliver the briefing on a tablet such as an iPad. The briefee would still get all the advantages of

DOI: 10.1057/9781137523792.0009

receiving a digitalized product but retain the ability to ask questions and generate immediate tasking. Some would argue that this process could also be digitized by adding a chat function that allowed for direct tasking of the analysts, but the author suspects this would almost certainly constitute a bridge too far for most managers – and also introduce some potential security issues. The inclusion of a chat or instant messaging function in a web-based analytic production process could also impose an unnecessary time burden on policymakers and decision makers, most of whom have a strong preference to task orally and engage in person. Good policy support almost always requires use of trusted interlocutors. The author's personal experience has been that policymakers who applaud the quality of the intelligence support they receive almost always go on to cite the person or persons who provide the service not the products themselves.

The primary rationale for moving to a new paradigm for delivering information and analysis is to leverage technology to support synchronous and asynchronous exchanges among the production team to improve the quality of the product. In recent years, web-based collaborative platforms that employ avatars have proven particularly effective in supporting synchronous dialogues among analysts and potentially between analysts and collectors as well as between analysts and policymakers (see Figure 4.3).[7] If a human connection has already been established between the producer and the consumer of analysis, then an avatar-based system could be used to supplant the daily briefing or, more important, to allow for conversations later in the day without requiring the briefer – or those who prepared the analysis – to travel to the consumer's office. Periodic in-person meetings would sustain the

FIGURE 4.3 *TH!NK Live™ avatar-based virtual world collaboration environment*

DOI: 10.1057/9781137523792.0009

personal relationship, but these sessions could be reserved for "deep dives" on a particular subject or limited to one a week or once a month.

Use of avatar-based collaboration platforms would facilitate more interaction and collaboration among analysts. In essence, analysts could replicate a coordination session without leaving their desks. This would instill more efficiency into the production process as it would remove a major barrier to collaboration – the cost of travel either along congested roads in a major city like Washington DC or across international boundaries. Use of the platform would also open new doors, allowing both analysts and collectors in far-flung locations across the world to work together as a team in generating the most complete and well-documented analysis possible.

Notes

1 *Wiki* is a Hawaiian word meaning "fast" or "quick." The first *wiki* was installed by Ward Cunningham on the Web in 1995, about a decade before *Facebook* and *Twitter* arrived on the scene. A good discussion of the role of *wikis* can be found in Audrey Watters, "Why *Wikis* Still Matter," *Edutopia: What Works in Education* (October 18, 2011) at http://edutopia.org/blog/wiki-classroom-audrey-waters.

2 The author recognizes that a *wiki*-based system that dealt with classified information would require additional controls and procedures, but that discussion extends beyond the scope of this paper.

3 Randolph H. Pherson and Richards J. Heuer, Jr., "Structured Analytic Techniques: A New Approach to Analysis," in Roger Z. George and James B. Bruce, eds, *Analyzing Intelligence: National Security Practitioners' Perspectives*, 2nd Edition (Washington, DC: Georgetown University Press 2014), p. 234.

4 A more extensive list of 11 questions has been developed, the "Getting Started Checklist" that can be found in Richards J. Heuer and Randolph H. Pherson, *Structured Analytic Techniques for Intelligence Analysis*, 2nd Edition (Washington, DC: CQ Press/Sage Publications, 2015), p. 47.

5 Randolph H. Pherson and Richards J. Heuer, Jr., "Structured Analytic Techniques: A New Approach to Analysis," in Roger Z. George and James B. Bruce, eds, *Analyzing Intelligence: National Security Practitioners' Perspectives*, 2nd Edition (Washington, DC: Georgetown University Press, 2014), p. 239.

6 Katherine Hibbs Pherson and Randolph H. Pherson, *Critical Thinking for Strategic Intelligence* (Washington DC: CQ Press/Sage Publications, 2013), p. 225.

DOI: 10.1057/9781137523792.0009

7 Globalytica, LLC has been using an avatar-based virtual world platform, TH!NK Live™, to support analytic instruction for several years with a high degree of success. Participants can master how the system operates in minutes and find it playful, engaging, and far more effective than video conferencing or other distributed e-learning platforms. For more information on the system and its capabilities, visit www.globalytica.com.

DOI: 10.1057/9781137523792.0009

5

Creating Impactful Intelligence: Communication Lessons from the Corporate Environment

Jonathan Calof

Abstract: *Actionable intelligence arises when the decision maker is convinced to execute on the recommendations provided in intelligence reports. This means that intelligence analysis must be communicated in a manner that will convince management to take action. This chapter summarizes lessons learned from a review of practitioner-oriented articles found in the Society of Competitive Intelligence Professionals (SCIP) literature (primarily* Competitive Intelligence Magazine *and* CI. Insight*), Global Intelligence Alliance (GIA) research, and conference papers from CI practitioner conferences. The objective is to better understand how the competitive intelligence field views the communication element of the intelligence process and, more importantly, how to make communications more effective. The author draws on his own experience as an intelligence researcher (academic) and consultant over the past 20 years.*

Keywords: communication; competitive intelligence; corporate analysis; decision maker impact; intelligence

Arcos, Rubén and Randolph H. Pherson, eds. *Intelligence Communication in the Digital Era: Transforming Security, Defence and Business.* Basingstoke: Palgrave Macmillan, 2015. DOI: 10.1057/9781137523792.0010.

 DOI: 10.1057/9781137523792.0010

Actionable intelligence arises when the decision maker is convinced to execute on the recommendations provided in intelligence reports. This means that intelligence analysis must be communicated in a manner that will convince management to take action. The shift toward presenting analysis in digital form as opposed to hardcopy products underscores the need to better understand how the field of competitive intelligence views the communication element of the intelligence process and, more importantly, how it proposes to make communications more effective. In this book, several authors have brought up technology-based communication solutions and other approaches based on their experiences and the relevant literature on this topic. This chapter helps the reader apply what they have read in previous chapters to develop their own contingency model for intelligence communication.

The importance of communication in competitive intelligence (CI)

Definitions of intelligence and discussions on intelligence most often include the word "actionable." What is the point of engaging in the planning, gathering, and assessing function of intelligence if the product does not help the decision maker take action? Often, one of the reasons that it is not actionable lies in it not being communicated properly.[1] Tim Kindler, who headed Kodak's intelligence unit wrote on communication that while "the three most important issues in real estate are location, location and location... the three most important things in CI are communication, communication, and communication."[2] In his article on "creating information that cannot be ignored" Fiora emphasizes how important the communication task is for competitive intelligence. He references a study done by Jim Collins called "From Good to Great" in which the author analyzed 11 high-performing companies. Fiora noted from the study that "great companies shared an ability to turn information into information that cannot be ignored." This he said was the "definition of actionable intelligence."[3]

The issue of intelligence being actionable has been at the forefront of competitive intelligence thought, and the role of communications has been at the center of this discussion. Intelligence communication is supposed to be a "catalyst for action."[4] No wonder that in a study looking at the competitive intelligence process, Calof and Miller found that

DOI: 10.1057/9781137523792.0010

upwards of 20% of time spent in the intelligence process was devoted to communications.[5]

Competitive intelligence communication practices. The current and future communication of competitive intelligence has been addressed in two studies done by (1) the Competitive Intelligence Foundation which looks at intelligence practice as of 2006 and (2) the GIA study published in 2010 that looks at intelligence practices in the year 2015 and beyond.

In 2006, the Competitive Intelligence Foundation published a study, "The State of the Art: Competitive Intelligence."[6] The study focused on the current state of competitive intelligence of SCIP (Society of Competitive Intelligence Professionals) members. The results were based on a survey of SCIP members from around the world. The study's findings provide a useful starting point for understanding how the field views the communications process. Figure 5.1 provides a breakdown of the communication vehicle and extent of use by the respondents.

The authors of the study note that a constant struggle has existed in the field to identify how to best deliver intelligence quickly and effectively. This is especially important because intelligence is being delivered to a wider and more diverse internal audience. Given this need, it was not surprising that respondents tended to use methods that were quick to use and could reach multiple people – 73% said that they used email frequently or sometimes. However, as will be discussed later in this chapter, the competitive intelligence literature suggests that presentations and staff meetings can be more effective ways of providing intelligence results

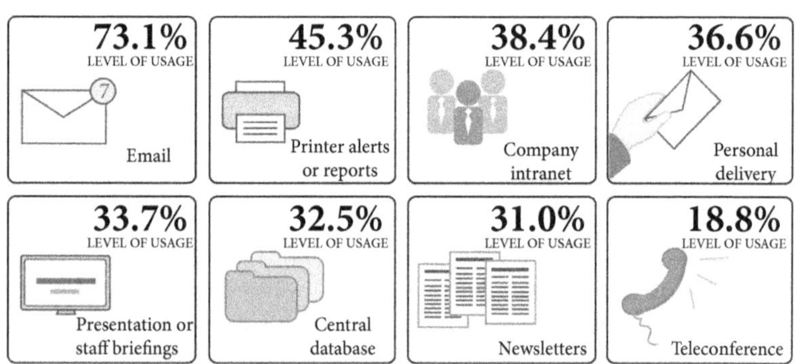

FIGURE 5.1 *Respondents' use of communication vehicles for disseminating intelligence*[7]

DOI: 10.1057/9781137523792.0010

that lead to action. Below is an excerpt from the report that provides a useful summary of the findings and the current thinking in competitive intelligence about intelligence communication.

> Presentations and staff meetings are among the most time-consuming and complex methods of delivering competitive analysis, yet they continue to be used by most CI practitioners and their use is correlated with management visibility. One-third of survey respondents use them *frequently*, and more than three-quarters do so *frequently* or *sometimes*.
>
> Presenting intelligence in front of a group or individual offers several advantages. It provides personal contact with clients, and affords the ability to engage in a discussion and receive immediate answers. New areas of interest can be quickly identified, along with the ability to quickly gauge reactions to the information or analysis presented. Depending on the skill of the presenter, it is often the most effective way to deliver complex or potentially controversial findings.
>
> The frequency of presentations seems important to survey respondents. Giving presentations or staff briefings is correlated with management visibility – the more frequent, the more likely there is also an increase in management visibility. *Rarely* or *never* delivering via presentations or staff meetings is often accompanied by less management visibility.

What do we learn from this study of communicating intelligence through the eyes of a competitive intelligence association?

1. A wide range of different mechanisms are used to deliver intelligence – eight are identified in this study.
2. Organizations use multiple communications methods (numbers add up to more than 100%).
3. Some types of dissemination vehicles provide more visibility with management (presentations being superior). The competitive intelligence field contends that visibility is an important concept, as will be discussed later in this chapter.
4. Interaction between the decision maker and the intelligence professional is desired.

The Global Intelligence Alliance study on Market Intelligence (MI) trends in 2015 and beyond[8] identified several key trends that would influence future intelligence practice. The term market intelligence was used by this organization rather than competitive intelligence but the core definition and function are similar. In fact, the authors acknowledge this in their discussion of definitions and terminology:

DOI: 10.1057/9781137523792.0010

> Market Intelligence (MI, frequently also used interchangeably with 'Competitive Intelligence, CI' or 'Business Intelligence, BI') is a distinct discipline by which organizations systematically gather and process information about their external operating environment (such as customers, competition, trends, regulation, or geographic areas). The purpose of Market Intelligence is to facilitate accurate and confident decision making that is based on carefully analyzed information.

The results are based on a survey conducted in 2010 with competitive/market intelligence professionals. Several communication-related aspects emerge from various elements in the study. For example, the study noted that changes in technology would lead to co-creation of intelligence. The communication implication is that "MI professionals will need to increasingly give briefings and presentations and engage in facilitating workshops such as scenario planning, war gaming, crowd forecasting, and trend seminars."[9] Under this approach, the intelligence professional is supposed to produce high level analysis and then present it in these forums to senior managers who then interact with the intelligence professional to produce recommendations. This concept of jointly producing intelligence through presentations and workshops can only arise if the intelligence professional is viewed by management as a trusted adviser and co-worker.

In terms of the actual outputs of the intelligence process, the GIA study noted that MI products in the future will be more sophisticated with increased analytical depth and variability in delivery. In particular, respondents envisioned intelligence outputs shifting from one time reports to online/just in time/frequently updated content. The online element is expected to grow more than face-to-face delivery and discussion. While this would appear to contradict the earlier findings in the study (joint creation) what it really says is that much of the co-creation will be done in virtual environments.

Another communication-related item noted in the study was that the future of intelligence communication would emphasize increased visualized intelligence deliverables, personalized delivery, and decision point intelligence. Figure 5.2 provides the GIA study findings in the areas of the process, deliverables, and tools in terms of trends to 2015.

The GIA study does provide support for the Fehringer et al. 2006 study in terms of multiple delivery mechanisms and engagement between the decision maker and the intelligence professional. Thus this must be seen as an important element in communication of intelligence. It also

DOI: 10.1057/9781137523792.0010

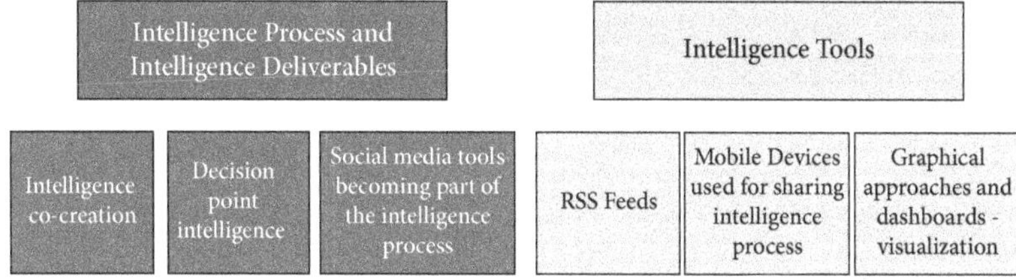

FIGURE 5.2 *Communication-related findings of Global Intelligence Alliance 2010 study*

provides other factors that need to be looked at in terms of the communication process, arguing that:

1 Decision maker interaction with the data and analysis will become more important.
2 Graphics will increasingly be valued over text.

The two studies offer a range of practices and considerations for the communication of intelligence and clear indications from the corporate environment on current and future communication approaches.

The Fehringer study also looked at different kinds of communication products (e.g., email, newsletters, reports, presentations). The study findings seem to suggest that companies use multiple methods to communicate intelligence. The GIA study notes the increasing number and sophistication of intelligence communication outputs. The SCIP literature, while favoring personal communications, does seem to suggest a mixed approach or multiple layers of communications.[10,11]

Key variables for effective communications

The previous section laid out various communication approaches including the idea of multiple communication outputs to increase the probability of acceptance of intelligence results. In this section, we examine the communication-related variables discussed in the competitive intelligence field that have impact on the acceptance of the intelligence results.

The first key variable is the extent of trust in the relationship between the person who communicates the intelligence and the customer of the intelligence. The GIA study argues that the concept of trust and the

DOI: 10.1057/9781137523792.0010

relationship between the decision maker and the intelligence professional is very important. Much of the SCIP literature also deals with the need to develop a trusted advisor relationship with the decision maker. Others have found that the quality of interaction at a personal level with the decision maker was related to whether the decision maker paid attention to the intelligence products.[12,13] Similarly Prescott and Williams noted that decision makers rely on those they trust: "A common frustration among competitive intelligence professionals is that their intelligence products are not completely made use of by decision makers. Because decision makers rely on those they trust, it is first necessary to establish trust between CI professionals and their users before CI will be fully utilized."

Trust and relationship are also cited as an important variable in the marketing and sales literature.[14,15,16] The author has personally observed the importance of this factor so many times in his consulting career that when he conducts intelligence projects for organizations, one of the first questions asked is: Who does the decision maker trust? Whoever is identified becomes a key person in the intelligence project and part of the intelligence communication team.

The job of the intelligence practitioner is to "sell" the intelligence results to the user of the intelligence. This has been a theme in many SCIP presentations. Selling becomes far easier when the "salesperson" is trusted by the decision maker.

When intelligence findings differ from decision maker beliefs

SCIP literature recognizes the difficulty in "selling" intelligence findings when a variance exists between what the intelligence is saying and the decision maker believes. Sullivan contends that people often have difficulty accepting information that is at variance with their beliefs.[17] Many of those in government, particularly from more traditional intelligence services, have similarly expressed frustration over being told what they are supposed to find and the frustration of having results ignored when they are at odds with the decision makers' beliefs.

Unfortunately, there is no easy fix for this problem. Research in this area points to the need for overwhelming evidence provided over extended periods of time to help change fundamental beliefs of decision makers.[18] However, the author's experience as well as research into trust

DOI: 10.1057/9781137523792.0010

and relationship suggest that the stronger the relationship between the decision maker and the intelligence creator the higher the probability of acceptance of intelligence findings and recommendations – even when they are at odds with decision maker beliefs.

Based on the growing evidence from practice and research on the impact of relationship and trust on "selling" intelligence recommendations, a critical need exists for the intelligence professional to establish a trusted relationship with decision makers. This importance may be related to the concern that in the business world many decision makers are not familiar with the intelligence function and process.[19] In the absence of this knowledge of the concept, it is not surprising that reliance on trust and relationship becomes more important.

Decision maker involvement in intelligence development

The traditional way of delivering intelligence was to provide a final, finished, and static report to the decision maker.[20] Any interaction with the decision maker was usually limited to initial meeting(s) to define intelligence needs and perhaps interim progress reports. The result is a static product.

Looking to the future, the trend is to move away from this approach. Two new elements of interaction are now emerging in the intelligence literature:

1 **Co-development of intelligence**: Seminars, workshops, meetings in which the intelligence provides high level analysis and then facilitates a discussion with the decision maker where the analysis is used to derive conclusions and actionable recommendations.
2 **Electronic delivery** that can be manipulated by the decision maker, providing her or him with the intelligence assessment and the ability to manipulate the data to develop additional insights. For example, IBM's Cognos system allows the decision maker to test other relationships within the data and explore the impact of other variables. If the data tables are in Excel, the decision makers could do a sensitivity analysis by changing the weights of variables.

Why is interaction with the client becoming so important? Why does it increasingly need to be part of the intelligence communication

DOI: 10.1057/9781137523792.0010

process? The answer is that these processes facilitate acceptance of the recommendations and planning for subsequent action. Decision makers are more inclined to accept recommendations if they are part of their development. Interaction by the clients with the data usually leads to higher acceptance of the results. As Klangsler wrote, "clients should be able to manipulate the illustrations contained in the research."[21] Hohhof similarly noted that "When analysts and their audience together build representations of issues, outcomes, and recommendations, the executives stay engaged and create their own "aha" moments which allow them to own the results. It's about creating a dialog between analysts and executives."[22]

This concept of interaction with the decision maker in the development and communication of the results is critical. Hohhof goes on to say "optimally the analyst spends the minimum time to present the intelligence and the maximum time engaging the executives in discussion. Dialogue creates acceptance, which leads to action."[23]

Interaction with the client in development of the intelligence and communication enhances learning, intelligence development, and intelligence acceptance, leading to action. However, as the learning literature has shown, only so many variables can be considered at the same time for this interaction effect to have its maximum impact. Past research has shown that the brain can process approximately seven items of information at a time.[24] This finding is at the core of cognitive load theory which posits that a limit exists regarding the number of variables that the decision maker can deal with at any one point in time.[25] As such, providing decision makers with a file in which they can manipulate the data has to be limited.

Decision maker learning styles

Many years ago, the author asked the head of intelligence at a large Canadian company what the best way was to communicate intelligence. His response, in essence, was that he had to present his findings to the senior management team (referred to as the "C" suite) and one of the officers likes long reports with appendixes and an Excel file that can be manipulated. Another officer, however, preferred a two-page summary, a third liked a seminar style presentation to her team, and a fourth wanted an oral briefing. Such variety in individual preferences in communication

DOI: 10.1057/9781137523792.0010

are not that unusual. In fact, the author has often stressed the need to do a profile on your own decision makers to learn how they make their decisions, style of communication they prefer.

Many individual decision maker factors need to be taken into consideration when determining the best method for communication. Myer's-Briggs is one of many psychological based tests that intelligence communicators should consider using. The Myers-Briggs distinction between sensing and intuition specifically addresses how the decision maker likes to receive information.

Cultural factors can also impact communications. Elizondo and Glitman pointed out,[26] in an article about delivering competitive intelligence to international audiences, how certain cultures prefer quantitative based presentations (Germans are an example of this) and others are more comfortable with qualitative based presentations (such as Americans).

Preference for graphics over text

A recurrent theme in the CI literature is the importance of increasing the use of graphics and the need to decrease the use of text. Global Intelligence Alliance in their study of the future of competitive intelligence pointed to "increasingly visualized intelligence deliverables ... graphs, dashboards and score cards."[27]

Dr. Johan Van Zyl, President and Chief Executive of Toyota South Africa, emphasized this in keynote address on competitive intelligence at a Knowledge Management conference.[28] At Toyota, they communicated intelligence with short deliverables only a few pages long that were filled mainly with graphics. He showed the audience the intelligence room at Toyota which had numerous graphics and tables on the wall and examples of intelligence products which made use of graphics including cartoons.

Why the emphasis on graphics over text? Why the focus on shorter reports? The SCIP literature provides several answers. In terms of the length of reports, much has been written about how busy executives are these days and their need to focus to capture their attention.[29,30] Learning theory posits that people learn better with a mix of graphics and text as opposed to text alone.[31] Kangsler argues that "pictures are better than words."[32] Similarly, Hohhof writes: "When possible, replace words with

DOI: 10.1057/9781137523792.0010

images" because learning theory contends that the primary purpose of graphics is to enhance learning.[33]

SCIP literature also points to an additional benefit, that of generating discussion which relates back to the value of interaction versus static presentations. As Fiora notes, "A relevant graphic or chart can spur a lively discussion as you and your audience discuss the implications of the data."[34]

Towards a contingency model of intelligence

The objective in intelligence communication is to develop the best approach for causing the decision maker to accept the intelligence recommendations and take action. What is the best way to present intelligence? The answer based on SCIP literature and the author's experience in intelligence consulting and research is that it depends on three broad sets of factors (see Figure 5.3):

1. The presenter's relationship with the decision maker and/or the extent to which the decision-maker trusts the person who developed/presented the intelligence.
2. The decision makes cognitive orientation and learning style.
3. The extent to which the intelligence developed is consistent or not consistent with the decision makers' beliefs (what they expected the intelligence results to be).

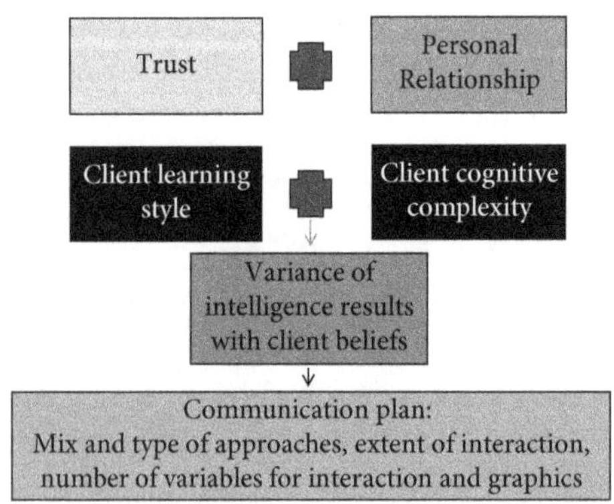

FIGURE 5.3 *Conditional communications and intelligence development model*

DOI: 10.1057/9781137523792.0010

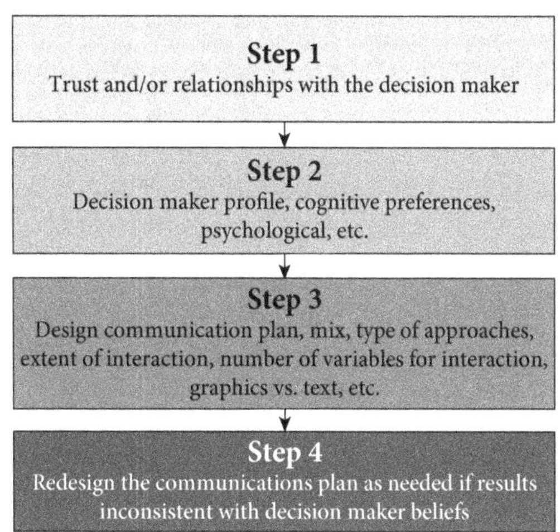

FIGURE 5.4 *The intelligence communication contingency Model*

Communication, therefore, is, and probably always will be, contingent on many factors; this chapter has provided three broad categories and multiple variables within each. Other variables can most likely be added to these categories as well. In all elements of the communications process, the decision maker profile needs to be taken into consideration. In particular, it is important to know how the recipients of intelligence like to receive analysis, who influences them, and what it takes to convince them to take action. This is a finding consistent with this chapter and the competitive intelligence literature.

The contingency model of intelligence communication integrates communication related decisions into the planning phase of intelligence. This four-step model for intelligence communications strategy is laid out in Figure 5.4 and described below.

Step 1: Establish if a strong relationship or trust exits between the decision maker and the person who will be communicating the intelligence. If one exists, then little additional thought has to be given to the communication element of intelligence. If this relationship or trust does not exist then identify a person who is trusted by the decision maker to present the intelligence.

Step 2: Profile the decision maker to identify his or her preferred method for receiving communication and factors that lead to their taking action on intelligence recommendation.

DOI: 10.1057/9781137523792.0010

Step 3: Design the communications strategy: Steps 1 and 2 will help determine the extent to which the decision maker needs to be involved in the co-creation of the intelligence. If they are to be involved, consider how many variables can they process at one time. The profile should also reveal the extent to which text vs graphics need to be used, cultural variables, and others that impact the messaging.

Step 4: Redesign the communications strategy: Step 3 is not a one-time exercise reserved for the beginning of project or activity. Rather, it should be an ongoing activity. For example, as the intelligence is being developed and the findings are deemed consistent with the decision maker's beliefs, then the relationship and trust factor becomes less relevant as does time and attention related to selling intelligence results based on the profile of the decision maker. Even active decision maker involvement in co-creation of the intelligence can be reduced.

On the other hand, if it becomes evident in the early stages of intelligence development that the results will not be consistent with the decision maker's beliefs, then Step 1 and Step 2 will need to be looked at again in the context of perhaps bringing in additional resources to "sell" the intelligence results. One organization brought in a well-known academic consultant to re-do their intelligence project and present it to their boss (the decision maker). It believed this was the only way the decision maker would accept the radical recommendations arising from their intelligence study.

Conclusion

Actionable intelligence arises when the decision maker is persuaded of the need to execute the intelligence recommendations. This requires that the intelligence must be communicated in a manner that will convince managers to take action. Appropriate attention also must be paid to items such as writing styles, organization, meaningful titles, and a focused message.[35] Each of the communication elements mentioned above arises from an author looking at ways to get the decision maker to act on the intelligence. The challenge is to determine how to convince the decision maker to accept the recommendations. A recurring theme in the research relates to the decision maker themselves, their preferences, their cognitive ability, their involvement in the development of intelligence, and the extent to which the intelligence professional is trusted.

DOI: 10.1057/9781137523792.0010

The first step in developing an appropriate communication strategy is to profile the decision maker to determine the best type of delivery vehicle, the extent of graphics and interaction, and the most appropriate style. Without this understanding, it will be difficult to execute most of the practices discussed in this book.

Notes

1 Calof, Jonathan. 'Communications and Trade Shows'. *Competitive Intelligence Magazine* 11(6), (2008): 43–45.
2 Kindler, Tim. 'Creating a Successful CI Operation in Today's Corporate Environment'. *Competitive Intelligence Magazine* 6(5), (2003): 6–9, p. 8 referenced.
3 Fiora, Bill. 'Creating Information That Cannot Be Ignored'. *Competitive Intelligence Magazine* 6(5), (2003): 36–37, p. 36 quoted.
4 Rothberg, Helen. 'Telling the Intelligence Story'. *Competitive Intelligence Magazine* 16(4), (2013).
5 Calof, Jonathan and Jerry Miller. 'Competitive Intelligence a Comparative Analysis'. *Proceedings, Society of Competitive Intelligence Professionals Annual Conference* (1997): 213, San Diego, CA.
6 Dale Fehringer, Bonnie Hohhof, and Ted Johnson. *State Of The Art Competitive Intelligence*. 1st ed. (Virginia: Competitive Intelligence Foundation, 2006).
7 *Ibid.*
8 Global Intelligence Alliance, *2010. MI Trends 2015: The Future of Market Intelligence*. White Paper March 2010 (Global Intelligence Alliance, 2010, p. 4).
9 Global Intelligence Alliance, *2010. MI Trends 2015: The Future of Market Intelligence*. White Paper March 2010 (Global Intelligence Alliance, 2010, p. 9).
10 Naylor, Ellen. 'Communicating Cooperatively'. *Competitive Intelligence Magazine*, 10(1), (2007): 44–46.
11 Kindler, Tim. 'Creating a Successful CI Operation in Today's Corporate Environment'. *Competitive Intelligence Magazine* 6(5), (2003): 6–9.
12 Lewis, Dafyyd. 'Rules Of Engagement: An Essential Prerequisite for Delivering Intelligence. Competitive Intelligence Visually'. *Competitive Intelligence Magazine* 6(5), (2003): 15–19.
13 Prescott, John, and Rachelle Williams. 'The User-Driven Competitive Intelligence Model: A New Paradigm for CI'. *Competitive Intelligence Magazine* 6(5), (2003): 10–14, p. 10 quoted.
14 Poon, Patrick, Gerald Albaum, and Peter Shiu-Fai Chan. 'Managing Trust in Direct Selling Relationships'. *Marketing Intelligence & Planning* 30(5), (2012): 588–603.

DOI: 10.1057/9781137523792.0010

15 Wood, John Andy, James S Boles, Wesley Johnston, and Danny Bellenger. 'Buyers' Trust of the Salesperson: An Item-Level Meta-Analysis'. *Journal of Personal Selling & Sales Management* 28(3), (2008): 263–283.
16 Young, Louise, and Gerald Albaum. 'Measurement of Trust in Salesperson – Customer Relationships in Direct Selling'. *Journal of Personal Selling & Sales Management* 23(3), (2003): 253–269.
17 Sullivan, Mark. 'Push Competitive Intelligence: Nested Internal Communication'. *Competitive Intelligence Magazine* 11(4), (2008): 10–12.
18 The growing use of structured analytic techniques may also help alleviate this problem in that they provide a transparent and systematic process for generating conclusion that the recipient of the analysis can track and critique. See Richards J. Heuer Jr. and Randolph H. Pherson, *Structured Analytic Techniques for Intelligence Analysis* 2nd Edition (Washington DC: CQ Press/SAGE Publications, 2nd Edition, 2015).
19 Calof, Jonathan. 'Competitive Intelligence Are We Really Becoming A Profession'. *Competitive Intelligence Magazine* 11(5), (2008): 16–20.
20 Global Intelligence Alliance. *Become A Trusted Advisor To Your CEO And The Top Brass*. Global Intelligence Alliance, 2012.
21 Kangiser, Angela. 'Delivering Competitive Intelligence Visually'. *Competitive Intelligence Magazine* 6(5), (2003): 20–23, referenced from p. 21
22 Hohhof, Bonnie. 'Communication Vs Presentation'. *SCIP Insight*, 2009.
23 Hohhof, Bonnie. 'Communication Vs Presentation'. *SCIP Insight*, 2009.
24 Miller, George A. 'The Magical Number Seven, Plus Or Minus Two: Some Limits On Our Capacity For Processing Information.' *Psychological Review* 63(2), (1956): 81.
25 Masri, Kamal, Drew Parker, and Andrew Gemino. 'Using Iconic Graphics in Entity-Relationship Diagrams: The Impact on Understanding'. *Journal of Database Management (JDM)* 19(3), (2008): 22–41.
26 Elizondo, Noe, and Erik Glitman. 'Delivering Competitive Intelligence to International Audiences'. *Competitive Intelligence Magazine* 6(3), (2003): 55.
27 Global Intelligence Alliance, *2010. MI Trends 2015: The Future of Market Intelligence*. White Paper March 2010. Global Intelligence Alliance, 2010 referenced from p. 13.
28 Van Zyl, Dr. Johan. 'Competitive Intelligence in the Knowledge Economy: A Reality or Just an Aspiration?' In *KCIM Conference*. Johannesburg, South Africa, 2012.
29 Sperger, Michael. 'May I Have Your Attention'. *Competitive Intelligence Magazine* 11(4), (2008): 13–16.
30 Rothberg, Helen. 'Telling the Intelligence Story'. *Competitive Intelligence Magazine* 16(4), (2013).
31 Masri, Kamal, Drew Parker, and Andrew Gemino. 'Using Iconic Graphics in Entity-Relationship Diagrams: The Impact on Understanding'. *Journal of Database Management (JDM)* 19(3), (2008): 22–41.

DOI: 10.1057/9781137523792.0010

32 Kangiser, Angela. 'Delivering Competitive Intelligence Visually'. *Competitive Intelligence Magazine* 6(5), (2003): 20–23.
33 Hohhof, Bonnie. 'Communication Vs Presentation'. *SCIP Insight*, 2009.
34 Fiora, Bill. 'Creating Information That Cannot Be Ignored'. *Competitive Intelligence Magazine* 6(5), (2003): 36–37, quoted from p. 36.
35 Rothberg, Helen. 'Telling the Intelligence Story'. *Competitive Intelligence Magazine* 16(4), (2013).

DOI: 10.1057/9781137523792.0010

6

Transforming Producer/ Consumer Relations through Modeling and Computation

Aaron B. Frank

Abstract: *Efforts to reform intelligence analysis have been motivated by the assumption that accurate analysis naturally leads to effective policy decisions. From this perspective, computational resources have primarily been devoted to the collection and assessment of empirical data in an effort to provide consumers with increasingly accurate predictions. By challenging the assumption that consumers welcome the predictions offered by intelligence analysts, a new perspective emerges regarding the ways in which the intelligence community may employ computational resources to develop increasingly useful and trusted analytic products and tradecraft. A new, model-centric analytic tradecraft that combines the computational resources of Big Data analysis, Agent-Based Models of artificial societies, and increasingly sophisticated and personalized human computer interaction technologies offer new opportunities to transform the relationship between intelligence producers and consumers.*

Keywords: agent-based modeling; analysis; computation; intelligence products; producer-consumer relations

Arcos, Rubén and Randolph H. Pherson, eds. *Intelligence Communication in the Digital Era: Transforming Security, Defence and Business.* Basingstoke: Palgrave Macmillan, 2015. DOI: 10.1057/9781137523792.0011.

A common assumption is that improvements in intelligence analysis naturally produce better decision making and policy. However, such an assumption rests on the belief that intelligence products are welcomed and used by consumers in political decision making processes. A more powerful measure for assessing the value of analysis is the Intelligence Community's (IC's) ability to provide more usable inputs into the political decision making process.

The IC is in need of finding a new tradecraft that is equally adept at navigating the politics associated with supporting multiple, often competing, stakeholders and having the agility to tailor its message to a wide variety of disparate customers. Before improvements in the predictive power or objectivity of analysis can be realized, analysts must first find ways to deliver assessments in more dynamic ways. Rather than view intelligence products as the output of collection and analysis, future reforms should reimagine intelligence products as inputs into a politicized policymaking process. Primacy should be given to questions of relevance and utility, rather than epistemologically problematic assessments of timeliness and accuracy, or long-standing debates regarding whether analysis can be more scientific or disciplined.[1]

The convergence of Big Data, Agent-Based Modeling (ABM), and Human-Computer Interaction (HCI) technologies provide a foundation upon which a new analytic tradecraft can be built. While this tradecraft would affect the IC's collection and analysis activities, its primary motivation is the development of new kinds of analytic products and relationships with consumers. Together, the combination of increasingly sophisticated computational capabilities should enable analysts to provide highly tailored, interactive analytic products that generate maps between analytic assumptions, data, policy options, and expectations of outcomes. Analysts then could offer multiple consumers a better understanding of high-dimensional tradeoff spaces that can assist stakeholders engaged in the politics of coalition formation and collective action in highly complex and uncertain situations.

The problem of producer-consumer relations

The relationship between intelligence analysts and policymakers is one of the most difficult problems facing scholars and practitioners. From the birth of the modern IC, producers have struggled to provide analysis

DOI: 10.1057/9781137523792.0011

that is relevant and useful to decision makers engaged in political and bureaucratic decision-making processes while remaining objective and preserving their institutional independence.[2] Three interrelated challenges provide a context that must be considered before any understanding of the ways in which new computational technologies might improve or aggravate relations between producers and consumers of intelligence. These three factors are: (1) the implications of prediction as the basis of intelligence support to policy; (2) the problems posed by deep uncertainty as a matter of epistemology; and (3) deep-seated institutional and cultural differences between intelligence and policy that establish competing definitions of support and roles that each should play.

Prediction and legitimacy in policy and intelligence

In an ideal model of producer-consumer relations, intelligence analysts provide consumers with predictions regarding the future state of the international system and policymakers welcome their input in the decision-making process. Intelligence analysts assist decision makers in identifying policy options and strategies with the highest likelihood of success, minimal cost, and lowest risk. To the extent policymakers benefit from these predictions and warnings of future events, improvements in analysis should naturally lead to better policy and decision making.

This model of producer-consumer relations is problematic. If one assumes that a policymaker's actions are determined by the predictions he or she receives, then this is incompatible with governance based on democratic representation, rule of law, and measures of political legitimacy. Because the power to make policy is legitimized by the process by which individuals are placed in their respective political and institutional roles, grounding producer-consumer relations in the provision of predictions by intelligence analysis introduces a new source of political authority that undermines the principles of democratic governance by bounding policy debates and choice-making. The use of intelligence estimates to set policy agendas and delimit or bind consumer's options and goals shifts the locus of policymaking to a class of technocratic experts who may possess highly-developed subject matter expertise but no ethical or institutional ties to constitutionally proscribed sources of legitimacy.[3]

Whatever role computational technologies and resources may play in the future of producer-consumer relations, the basis of this relationship

DOI: 10.1057/9781137523792.0011

is unlikely to be grounded on providing predictions alone, even if these capabilities allow for analysts to offer increasingly accurate and reliable knowledge about the future.

Deep uncertainty, epistemology, and decision-making

A secondary problem with prediction is that it promotes a deterministic vision of the international system and human affairs. Such a view is ironic given that a predictable world is necessarily one where the future cannot be changed by the choices and actions of decision makers – either because human agency offers no relief from structural forces in the shaping of social systems or because actors cannot choose to protect their private information such as plans, intentions and capabilities, meaning that no uncertainties exist. A better context is required for characterizing how consumers make choices that are sensitive to epistemological limitations of what can be credibly predicted.

Policymaking occurs within a framework of deep uncertainty.[4] Deep uncertainty occurs when producers and consumers cannot agree upon an appropriate conceptual model for identifying and characterizing relationships between forces that will shape the future, the probability distributions used to represent uncertainties around key variables, and how alternative outcomes should be valued, ranked, or otherwise assessed.[5] Deep uncertainty affects the intelligence process in three important ways, each of which can exacerbate tensions between producers and consumers.

- Deep uncertainty allows for and encourages the use of multiple competing perspectives or models of strategic problems, each of which provide a legitimate basis for analysis and decision-making. As a result, consumers may possess distinct and divergent expectations about the future of the international system resulting in alternative frames against which intelligence products may be evaluated.
- The presence of multiple, credible frameworks for characterizing strategic problems means that empirical information is often ineffective at resolving policy and analytic disputes. The available data may be consistent with multiple competing models, may be irrelevant to one or more competing perspective, or may be interpreted in many different ways further expanding, rather than reducing, the range of uncertainties producers and consumers must consider.[6]

DOI: 10.1057/9781137523792.0011

- In the context of bureaucratic politics, the presence of assessments that support the expectations derived from one framework while militating against others are inherently viewed as political, even if the assessment's source operates independently of the political process.[7] Thus, in cases where governance requires stakeholders to form coalitions in order to set agendas, define authorities, and mobilize resources, policy and politics become inseparable, and all information becomes politicized.

When viewed through the IC's foundational epistemology that divides intelligence into basic, current, and estimative – verifiable facts, reporting on current events, and speculations about missing data or possible futures – the challenges posed by deep uncertainty are apparent (see Figure 6.1).[8] As Kent, Hilsman, and others have noted, consumers generally welcome basic intelligence products into the decision-making process, often because few consumers have the time or expertise to become familiar with all of the relevant factual or historical knowledge that IC experts possess.[9] As analytic products shift from verifiable facts to descriptions of the current behavior of

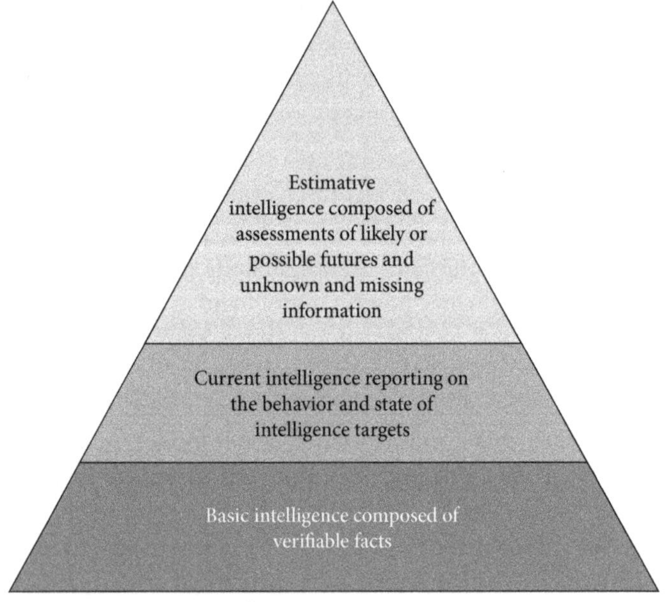

FIGURE 6.1 *Sherman Kent's intelligence pyramid*

DOI: 10.1057/9781137523792.0011

intelligence targets, relations between producers and consumers become increasingly conditional due to the increased ability of decision makers (and their staff) to perform their own assessments, often based on personal contact and relations with intelligence targets and experts outside of the IC.[10]

Finally, because estimates rest upon analysts' employment of theory and judgment regarding available facts, observations, and informed speculation about what intelligence targets might do, what options they have, and what their reactions to others' policies or actions could be, consumers are more inclined to challenge the validity of the producers' assessments.[11] Thus, a paradox exists regarding the ability of intelligence producers to offer predictions or otherwise reduce uncertainty for consumers because estimates regarding the future state of the international system and the success or failure of given policies are the most likely to be rejected on the grounds that they venture beyond the world of known facts and observations.[12]

The presence of deep uncertainty and the epistemological paradox posed by consumers' concerns over intelligence assessments that venture beyond the limitations of empirical verification and observation provide an initial context within which computational resources should be applied in order to improve the relevance of intelligence products.

Institutional problems

The institutional arrangement between producers and consumers places intelligence analysts in a position that is simultaneously characterized by independence and subordination. As a result, how closely analysts and policymakers should work together remains an unsolved and perpetual problem where tradeoffs entail the choice between politicization and irrelevance.[13]

While studies of politicization suggest that producers rarely apply direct pressure to analysts and analysts rarely engage in blatant pandering, the privileged position of consumers in their relationship with intelligence producers affords several opportunities to shape analysis through institutional prerogatives, such as calling for independent assessments, creating new analytic units, questioning analysts about specific data sources and methods, setting analytic and collection priorities, or even threatening producers with the loss of access to stakeholders.[14]

DOI: 10.1057/9781137523792.0011

Institutional problems in the relationships between producers and consumers stem from the competing expectations regarding the role of intelligence professionals in the policy process, which has been characterized as competing tribes of optimists and pessimists.[15] In this model, consumers are fundamentally activists committed to shaping the world and events within it, and "... approach problems with the belief that they can solve them. After all, this is the reason they have gone into government."[16] Their optimism is an expression of agency and empowerment based on their position in a competitive, strategic international system that constantly begs attention and action. As a result, they believe in the importance of their choices and seek input from those whom they believe are committed to their success. In this context, bureaucratic and organizational politics have real consequences because they grant or deny decision makers access to resources and the authority to act, thus incentivizing the competition for control over agendas and information.[17]

Consumers rarely view intelligence products as neutral. Instead, intelligence products are seen as weapons to be wielded within the context of bureaucratic politics, agenda setting, coalition formation, and resource mobilization. Thus, rather than seeking analysis that challenges assumptions, raises nuance, and presents alternative frameworks through which strategic situations can be assessed, consumers often adopt a position of "accept or reject" towards intelligence products based on whether they advance or hinder their standing with other stakeholders.[18]

While consumers are regarded as optimists, producers are often considered pessimists whose institutional and professional orientation is focused on the development of deep, substantive expertise, an appreciation for the nuance, detail, and complexity of intelligence issues, policy neutrality, and seek to contribute to policy by warning decision makers when their strategies and policies may rest on unwarranted assumptions or be overtaken by events.[19] Consumers often believe that intelligence analysts are overly concerned with examinations of how their policies may fail, and not invested in the identification of opportunities to aid in their success.[20]

The primacy producers assign to sources of policy failures and vulnerabilities place them in a competitive, even antagonistic position with consumers who often resist analysis that suggests the results of their hard-fought bureaucratic battles will fail. While intelligence analysts believe that their skepticism protects policymakers from overconfidence and wishful thinking, consumers often perceive their warnings as a form

DOI: 10.1057/9781137523792.0011

of criticism and a challenge to their judgment and goals.[21] By constantly warning of failures and threats, intelligence analysts are seen as the bearers of bad news who are not team players at best, and actively working against the interests of their consumers at worst.[22]

These tensions are further exacerbated by the timing of decision-making and the need to act vs. the demand for more or better information. Policymakers believe that they do not have the luxury of waiting for more information or doing nothing as a crisis looms and must choose between many bad or sub-optimal options.[23] Consumers understand that they need to act on unfavorable terms and learn, adapt, and evolve solutions and manage crises that cannot be solved.[24] In such circumstances, consumers demand insights into the relative merits of available options, rather than simply accounting for the downsides of each.[25]

Because consumers believe that their choices matter, the ways in which they think about problems, define success and failure, and evaluate options become central concerns for mobilizing and sustaining support for their agendas and policies. Thus, the policy world is primarily and necessarily a world of ideas that are focused on achieving or avoiding future possibilities rather than facts and data about the past and present. Given the privileged position of consumers' ideas and beliefs in the decision-making process as well as their beliefs about agency, computational resources must offer producers more than a new means for providing data or identifying vulnerabilities in plans and strategies. They should also aid in the exploration of ideas about the how the international system operates and is influenced by individual and collective action.

The concerns over generating and receiving predictions permeate the relationship between producers and consumers, creating a context in which computational resources are employed in the development of policy and decision-making. Challenges posed by democratic governance, deep uncertainty, and beliefs about agency all indicate that improving the relevance of intelligence analysis as an input into political decision-making may be as, if not more, important than increased predictive accuracy.

The emerging computational landscape

The computational power available to producers and consumers has reached unprecedented levels and continues to expand. The ability to

DOI: 10.1057/9781137523792.0011

collect and analyze data at scales and speeds previously unimagined has created opportunities and challenges. Big Data, bulk collection, digital exhaust, the internet of things, cloud computing, and many other terms have emerged in recent years, redefining the ways computers are employed in intelligence collection and analysis. However, the IC has yet to capitalize on the transformative potential of Big Data, broadly regarded as the ability to aggregate and analyze the contents of massive, disparate data sets, ABM or artificial societies, and HCI, inclusive of the advent of secure, mobile communication and data processing capabilities.

Big Data and intelligence analysis

To date, the challenges and opportunities associated with Big Data have largely focused on the empirical aspects of intelligence collection and analysis, and the potential to rationalize decision-making by drawing upon a more complete and deeper body of evidence. As a result, debates have focused on matters of information access, security, and privacy, as exemplified by concerns over the bulk collection and assertions of the "right to be forgotten," which proceed from the assumption that data is valuable and can be exploited in a fashion relevant to decision makers' needs.[26]

Big Data computing capabilities have discovered patterns that were too complex to identify in smaller, isolated datasets and enabled evidence-based decision-making in a variety of fields such as medicine, traffic management, energy, marketing, and finance. In each of these cases, benefits arise from gaining a more complete description of the system under examination, improved classification of normal and outlying cases, or the ability to predict future outcomes based on projecting historical trends and established relationships into the future.

Efforts to introduce Big Data into the IC have focused on the increased development of predictive intelligence, anticipatory intelligence, Activity-Based Intelligence, and other terms that all rest on the extrapolation of future conditions from patterns identified in empirical data.[27] However, given the central role of agency in the international system, there are limitations to what Big Data can provide senior policymakers when compared with those gains being found in other domains.

- A reliance on Big Data assumes that observed cases and patterns will persist into the future and privilege those variables and conditions that recur consistently over those that are idiosyncratic.

DOI: 10.1057/9781137523792.0011

The implicit assumption that the future of social systems can be predicted based on their prior history antagonizes consumers by biasing structure over agency. It also diminishes the consideration of decision maker's choices and their ability to alter the course of history.[28] Likewise, Big Data, and induction more broadly, extracts information from populations of cases without focusing on specific instances, yet consumers deal with singular cases and accounts that may benefit from broad knowledge about populations of cases but demand detailed attention to the nuance of specific circumstances in particular instances.[29]

- In cases where reliable patterns can be identified in Big Data, organizational procedures and technological automation will exploit these patterns and remove them from the agenda of items requiring the time, attention, and judgment of senior decision makers. As a result, the long-term implications posed by discoveries in Big Data will be adverse selection based on the complexity of problems and the extent to which they are capable of generating continuous novelty based on individual or collective action.
- The classes of problems that consumers confront often extend beyond the limitations of the empirical record and demand a search across multiple counterfactuals and future scenarios. In these instances, decision makers may be confronted by no data rather than Big Data, thus shifting the basis of their choices from the available evidence to theory, heuristics, and social processes that determine whether particular choices can be justified or defended from future critiques informed by hindsight.[30]

Agent-based modeling and social simulation

The IC has largely eschewed the application of formal models in intelligence production. ABMs provide a means for adding rigor to analysis without imposing many of the constraints of formal mathematical models that have impeded the use of formal models in the use of decision-support.[31] Much like the popular games *The Sims* and *Spore*, ABMs are composed of heterogeneous, autonomous actors that can interact with one another and their environment based on any set of behavioral and learning rules that can be represented algorithmically.[32] As a result, a diverse range of theory and heuristics regarding human

DOI: 10.1057/9781137523792.0011

and organizational decision-making and behavior can be simulated computationally, allowing for intelligence producers and consumers to examine alternative descriptions of intelligence problems through the analysis of artificial societies composed of interacting agents in software.

ABMs have assisted in the development of new theories and testing of hypotheses whenever one or more of the following properties are present:

- Agents in the system are **heterogeneous** with respect to their attributes and/or behavior, and cannot be represented as an average or aggregate unit.
- Agents are **autonomous** in the sense that they are each capable of making decisions and acting according to their individual goals, capabilities, and information and are not controlled by a central authority.
- Agents reside in **explicit space** such as geographic terrain, social networks, or other abstract environments in which notions of distance, closeness, and locality affect interactions and experiences.
- Agents' behavior is **boundedly rational**, heuristic, and constrained by cognitive limitations.
- Model developers and users are concerned with the **non-equilibrium dynamics** of systems such as phase transitions, transients, tipping or branching-points, cycling, and other properties that may not be observed through the comparative statics.[33]

In the context of intelligence analysis, ABMs provide a basis for grounding estimative intelligence in synthetically generated data derived from specified individual and collective behavior. In doing so, the epistemological basis of intelligence analysis can be expanded, helping hedge against consumer's distrust of estimates by grounding assessments in simulated data whenever empirical information is missing or contested. Because ABMs afford great flexibility regarding the representation of individual and group behaviors, alternative competing ideas can be represented, examined, and compared via simulation, allowing for analysts to experiment with social systems *in-silico* in order to identify tradeoffs between alternative frameworks or descriptions of intelligence problems.[34]

DOI: 10.1057/9781137523792.0011

Interactive computing technology

Technology has made important in-roads into the producer-consumer relationship, as demonstrated by the delivery of the President's Daily Brief on an iPad.[35] Cloud computing, secure mobile wireless technology, video conferencing, touchscreen control, and software that learn users' preferences and behaviors provide an opportunity to offer consumers new kinds of analytic products and alter the relationship between producers and consumers.

Efforts to employ new technologies in producer-consumer relations have largely concentrated on making intelligence products more accessible to consumers. Additionally, technologies have been employed to enable consumers with opportunities to provide feedback to analysts, such as highlighting materials of interest, judgments that require greater elaboration, and the articulations of follow-on questions or lines of inquiry.[36] As the preceding chapters in this book have demonstrated, new technologies can enable the production of tailored analytic products that employ interactive multimedia in order to communicate with consumers.

The more challenging implications of a shifting technological base for producer-consumer relations involve identifying how to bring consumers into the intelligence production process without politicizing the IC's analytic processes and products.[37] For example, the advent of secure, mobile wireless technology allows for relatively light-weight computational devices to access and operate on massive quantities of data without ever moving sensitive information onto the device itself. This could enable the dissemination of intelligence in locations and to users in situations where it is currently too cumbersome to access, broadening the base of consumers while allowing decision makers increasing access to raw collection, working products, and individual analysts. While producers have often resisted allowing consumers unfettered access to intelligence information and personnel, it seems unlikely that these barriers can be maintained if priority is given to analytic transparency in an effort to improve the relevance and usability of assessments.[38]

Increasing interactivity between producers and consumers suggests additional applications of computational resources in their relationship. For example, learning systems may assist producers in understanding the criteria used by consumers to evaluate the relevance or accuracy of intelligence assessments, and offer insights into the ways in which

DOI: 10.1057/9781137523792.0011

decision makers have framed problems in order to aid in the tailoring of analytic products. When combined with additional applications of computational resources, such as Big Data and the creation and examination of artificial societies, a larger set of possibilities regarding how computational resources might be employed in the context of producer-consumer relations can be seen.

Developing a model-centric analytic tradecraft

A model-centric analytic tradecraft offers new opportunities to address many of the long-standing, fundamental problems in producer-consumer relations by capitalizing on increasingly powerful and available computational resources. This new tradecraft would extend the IC's ongoing efforts to institutionalize Structured Analytic Techniques (SATs), which have already established strong linkages between analysis, modeling, collaboration, and presentation.[39] By emphasizing the development and assessment of computational models as vehicles for exploring alternative ideas, data generation, and communication within the IC and with consumers, this new tradecraft shifts the emphasis of intelligence production from predictive accuracy and institutional independence to exploratory assessments of intelligence issues developed in collaboration with consumers.[40]

This model-centric analytic tradecraft employs computation to increase the relevance of intelligence analysis by addressing sources of tension in producer-consumer relations:

- By emphasizing the development and examination of ABMs, analysts can provide assessments that privilege the role of decision-making and agency in social systems. Moreover, by relying on computational models that express the behaviors of actors of the international systems as algorithms rather than equations, producers can represent a wide range of beliefs, frameworks, heuristics, or mental models formally, including those of consumers.
- By using formal models to generate data about alternative futures, analysts can shift the epistemological basis of intelligence estimates from one of informed speculation to inferences made from synthetic data produced by artificial societies. Such a transition not only places analysts on firmer ground when presenting and

DOI: 10.1057/9781137523792.0011

justifying to skeptical consumers, but also enables the application of Big Data computational resources in the generation and mining of simulation data.

- Increasingly sophisticated HCI capabilities can enable consumers to interact directly with massive quantities of data through the use of secure, mobile, interactive computing technologies. These capabilities allow for intelligence assessments to achieve the higher levels of accessibility across the community of established and potential consumers. They also allow assessments to be increasingly tailored with respect to how information is visualized, the sequence by which it is presented, and the kinds of narratives that are employed to explain the processes by which data has been collected, generated, and otherwise evaluated.

Together, the combination of these computational capabilities can pave the way for new kinds of analytic products to address challenges associated with deep uncertainty, agency, and interactivity. For example, by employing ABMs as tools for formalizing the mental models of consumers, in addition to those of analysts (and their beliefs about intelligence targets), producers can bring multiple stakeholders into the analytic process. Moreover, by using multiple models, each representative of alternative perspectives on intelligence problems, deep uncertainty can be addressed by providing a rich epistemological assessment of the robustness of particular analytic judgments regarding the likelihood of potential futures and the efficacy of alternative policy actions.

Figure 6.2 provides a notional example of a model ensemble, examining three alternative data sets that define initial modeling parameters, three alternative models that describe the target system in different ways, and three prospective policy actions. When all 27 combinations are considered, a tradeoff space is created where intelligence producers and consumers can evaluate the expected outcome of each combination of data, model, and action against multiple measures of performance, shown by the letters A–E. Analytic products of this, and other types, serve to provide consumers with maps that link alternative beliefs about the state of the world, how it works, and their options for action. Such maps help them better understand sources of disagreement regarding the valuation of options and outcomes as well as locate sets of data, theory, and options that would best explain the observed or expected behavior of intelligence targets.

DOI: 10.1057/9781137523792.0011

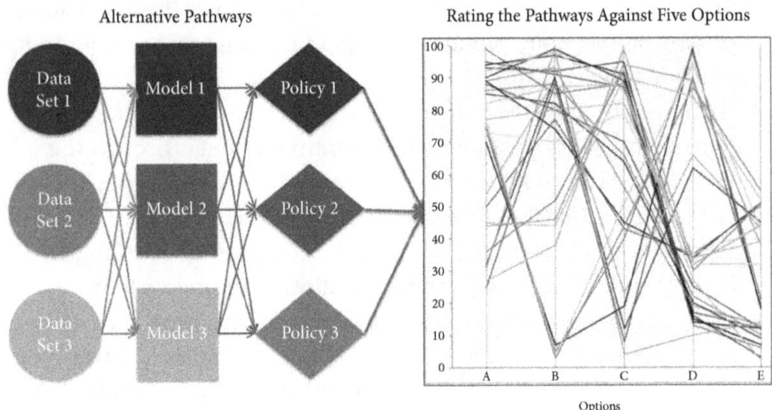

FIGURE 6.2 *Understanding the tradeoffs involved in implementing different combinations of data sets, models, and policy choices*

By identifying classes of outcomes that may be robust across many alternative model formulations or parameter settings, producers can offer consumers information that may assist in the formation of coalitions and constituencies for particular policy actions and metrics for their evaluation. They can also assist in socializing the merits or concerns associated with desired actions – all needed to effectively support decision making in the inherently politicized policymaking process. Moreover, troubling outcomes that may only be reached based on a series of contingent or special circumstances, and other points in which expectations differ based on changes in a handful of key assumptions, can be dealt with more efficiently.

Conclusion

Computational technology and resources are changing the relationship between intelligence producers and consumers. As new capabilities allow for increasingly tailored and interactive forms of presenting analysis, a larger opportunity exists for transforming analytic tradecraft more broadly and directly addressing fundamental problems in the producer-consumer relationship. While multiple efforts at reforming the IC have sought to address organizational and operational challenges associated with increasingly complex needs for collection and collaboration, implicit and naïve assumption about analysis and prediction that must be changed. Focusing on how to make intelligence more

DOI: 10.1057/9781137523792.0011

predictive instead of on how to make it more useful to decision makers and the decision-making process may exacerbate tensions in producer-consumers relations rather than alleviate them and leave the IC in worse position vis-à-vis those whom they support. Instead, the IC should seek to consolidate and promote lessons learned regarding the communication of intelligence analysis with decision makers and develop new tradecraft devoted to making analysis more relevant to the needs of stakeholders engaged in political decision-making under deep uncertainty.

Building a museum of analysis

Intelligence production resulting from a model-centric analytic tradecraft may be considered as the building of a museum in which each model of the ensemble constitutes a particular exhibit. Consumers are free to enter and exit the museum from any point they wish, and can tour the exhibits according their interests. Analysts act as docents who explain how each exhibit was developed and its relationships to other parts of the museum, e.g. a related exhibit arrives at similar conclusions using a model that employed different assumptions about a target's behavior and decision-making style.

While the museum allows for consumers to seek analysis that is attractive to them, whether based on their comfort with particular assumptions or interest in certain outcomes, it is also a collective and integrated product in its own right. Thus, efforts to cherry pick or only cite those exhibits that support a desired policy agenda can be contextualized by other stakeholders who are aware of the entire contents of the museum and all of the exhibits within it.

Finally, by tracking how consumers move about the museum, analysts can determine if its holdings are complete, if certain exhibits are being avoided, and develop new products to address unmet analytic demands or create new linkages between exhibits to ensure that consumers are aware of those exhibits that might be ignored in the policy debate yet merit consideration.

Notes

1 Walter Lacquer, *A World of Secrets: The Uses and Limits of Intelligence* (New York, NY: Basic Books, 1985); Kerbel, Josh. "Lost for Words: The Intelligence Community's Struggle to Find its Voice," *Parameters* (Summer 2008),

pp. 102–112; Jennifer E. Sims, "Decision Advantage and the Nature of Intelligence Analysis," in Loch K. Johnson, ed., *The Oxford Handbook of National Security Intelligence* (New York, NY: Oxford University Press, 2010), pp. 389–403; Richards J. Heuer, Jr., "The Evolution of Structured Analytic Techniques," *Presentation to the National Academy of Science, National Research Council Committee on Behavioral and Social Science Research to Improve Intelligence Analysis for National Security* (Washington, DC December 8, 2009), available at: https://www.e-education.psu.edu/drupal6/files/sgam/DNI_Heuer_Text.pdf (accessed on September 15, 2014); and Marrin, Stephen. "Is Intelligence Analysis an Art or a Science?" *International Journal of Intelligence and Counterintelligence* Vol. 25(3), (2012): 529–545.

2 Sherman Kent, *Strategic Intelligence for American World Policy* (Princeton, NJ: Princeton University Press, 1949), p. 180; Paul R. Pillar, "The Perils of Politicization," in Loch K. Johnson, ed., *The Oxford Handbook of National Security Intelligence* (New York, NY: Oxford University Press, 2010), pp. 472–484; and Katherine Hibbs Pherson and Randolph H. Pherson, *Critical Thinking for Strategic Intelligence* (Washington DC: CQ Press/SAGE Publications, 2013), pp. 159–170 (chapter 14, "How Do I Deal With Politicization?").

3 Mark M. Lowenthal, "The Policymaker-Intelligence Relationship," in Loch K. Johnson, ed., *The Oxford Handbook of National Security Intelligence* (New York, NY: Oxford University Press, 2010), p. 450.

4 Robert J. Lempert, Steven W. Popper, and Steven C. Bankes, *Shaping the Next One Hundred Years: New Methods for Long-Term Quantitative Policy Analysis* (Santa Monica, CA: RAND, 2003), p. xii.

5 Similar characterizations of uncertainty exist such as Knightian uncertainty, unknown vs. unknowable uncertainties, and ignorance. See Frank H. Knight, *Risk, Uncertainty and Profit* (Mineola, New York: Dover Publications, 2006); and Richard J. Zeckhauser, "Investing in the Unknown and Unknowable," in Francis X. Diebold, Neil A. Doherty and Richard J. Herring, eds., *The Known, the Unknown, and the Unknowable in Financial Risk Management* (Princeton, NJ: Princeton University Press, 2010), pp. 304–343.

6 Richards J. Heuer, Jr., *The Psychology of Intelligence Analysis* (Reston, VA: Pherson Associates, 2007); and Richards J. Heuer, Jr. and Randolph H. Pherson, *Structured Analytic Techniques for Intelligence Analysis* 2nd Edition (Washington, DC: CQ Press/SAGE Publications, 2015), pp. 181–192.

7 Paul Wolfowitz, "Comments: Paul Wolfowitz," in Roy Godson, Ernest R. May and Gary Schmitt, eds., *U.S. Intelligence at the Crossroads: Agendas for Reform* (Washington, DC: Brassey's, 1995), pp. 75–80.

8 Sherman Kent, *Strategic Intelligence for American World Policy* (Princeton, NJ: Princeton University Press, 1949), p. 11–29; and Kent, Sherman. "Estimates and Influence". *Studies in Intelligence* 12(3), (1968): 14–17.

DOI: 10.1057/9781137523792.0011

9 Hilsman, Roger. "Intelligence and Policy-Making in Foreign Affairs," *World Politics* 5(1), (October 1952):1–45; and Sherman Kent, "Estimates and Influence," *Studies in Intelligence* (Summer 1968), available at: https://www.cia.gov/library/center-for-the-study-of-intelligence/csi-publications/books-and-monographs/sherman-kent-and-the-board-of-national-estimates-collected-essays/4estimates.html (Accessed on: September 16, 2014).

10 Sherman Kent, *Strategic Intelligence for American World Policy* (Princeton, NJ: Princeton University Press, 1949), p. 38; and Kent, Sherman. "Estimates and Influence". *Studies in Intelligence* 12(3), (1968), p. 15; and James B. Steinberg, "The Policymaker's Perspective: Transparency and Partnership," in Roger Z. George and James B. Bruce, eds. *Analyzing Intelligence: Origins, Obstacles, and Innovations* (Washington, DC: Georgetown University Press, 2008), p. 84.

11 Sherman Kent, *Strategic Intelligence for American World Policy* (Princeton, NJ: Princeton University Press, 1949), pp. 40, 45–46; and Richards J. Heuer, Jr., "The Evolution of Structured Analytic Techniques," *Presentation to the National Academy of Science, National Research Council Committee on Behavioral and Social Science Research to Improve Intelligence Analysis for National Security* (Washington, DC December 8, 2009), available at: https://www.e-education.psu.edu/drupal6/files/sgam/DNI_Heuer_Text.pdf (accessed on: September 15, 2014).

12 Roger Z. George, "Central Intelligence Agency: The President's Own," in Roger Z. George and Harvey Rishikof, eds., *The National Security Enterprise: Navigating the Labyrinth* (Washington, DC: Georgetown University Press, 2011), p. 163.

13 Wirtz, James J. "Intelligence to Please? The Order of Battle Controversy During the Vietnam War". *Political Science Quarterly* 106(2), (Summer, 1991): 239–263; Richard K. Betts, *Enemies of Intelligence: Knowledge and Power in American National Security* (New York, NY: Columbia University Press, 2007), pp. 66–103; Gregory F. Treverton, "Intelligence Analysis: Between "Politicization" and Irrelevance," in Roger Z. George and James B. Bruce, eds. *Analyzing Intelligence: Origins, Obstacles, and Innovations* (Washington, DC: Georgetown University Press, 2008), p. 92; Kerbel, Josh and Olcott, Anthony. "Synthesizing with Clients, Not Analyzing for Customers". *Studies in Intelligence* 54(4), (December 2010):11–27; and interview with Paul Pillar, Georgetown University, February 2, 2012.

14 For discussions see Wirtz, James J. "Intelligence to Please? The Order of Battle Controversy During the Vietnam War". *Political Science Quarterly* 106(2), (Summer 1991):239–263; Gregory F. Treverton, *Reshaping National Intelligence for an Age of Information* (New York, NY: Cambridge University Press, 2003), p. 198; Pillar, Paul R. "Intelligence, Policy, and the War in Iraq". *Foreign Affairs* 85(2), (March/April 2006):15–27; Pillar, Paul R. "The Right Stuff". *The National Interest* No. 91 (September/October 2007): 53–59; Paul R. Pillar, "The Perils of Politicization," in Lock K. Johnson, ed., *The Oxford*

DOI: 10.1057/9781137523792.0011

Handbook of National Security Intelligence (New York, NY: Oxford University Press, 2010), p. 477; Paul R. Pillar, *Intelligence and U.S. Foreign Policy: Iraq, 9/11, and Misguided Reform* (New York, NY: Columbia University Press, 2011); and Joshua Rovner, *Fixing the Facts: National Security and the Politics of Intelligence* (Ithaca, NY: Cornell University Press, 2011).

15 Lowenthal, Mark M. "Tribal Tongues: Intelligence Consumers, Intelligence Producers". *The Washington Quarterly* 15(1), (1992): 157–168.

16 Mark M. Lowenthal, *Intelligence: From Secrets to Policy* (Washington, DC: Congressional Quarterly Press, 2009), p. 186.

17 John McLaughlin, "Serving the National Policymaker," in Roger Z. George and James B. Bruce, eds., *Analyzing Intelligence: Origins, Obstacles, and Innovations* (Washington, DC: Georgetown University Press, 2008), pp. 71–72.

18 Nearly a decade after characterizing his view of intelligence products, Wolfowitz played a pivotal role in controversial use of intelligence information during the run up the invasion of Iraq in 2003. During this period, he noted the central importance of intelligence assessments characterizing Iraqi WMD capabilities, not because of their factual content, but due to their central role in mobilizing government action by creating a for action. Paul Wolfowitz, "Comments: Paul Wolfowitz," in Roy Godson, Ernest R. May and Gary Schmitt, eds., *U.S. Intelligence at the Crossroads: Agendas for Reform* (Washington, DC: Brassey's, 1995), p. 76; and Paul Wolfowitz, *Deputy Secretary Wolfowitz Interview with Sam Tannenhaus, Vanity Fair* (May 9, 2003) http://www.defense.gov/transcripts/transcript.aspx?transcriptid=2594 (Accessed on May 30, 2012).

19 James B. Bruce and Roger Z. George, "Intelligence Analysis — The Emergence of a Discipline," in Roger Z. George and James B. Bruce, eds. *Analyzing Intelligence: Origins, Obstacles, and Innovations* (Washington, DC: Georgetown University Press, 2008), p. 9.

20 John McLaughlin, "Serving the National Policymaker," in Roger Z. George and James B. Bruce, eds., *Analyzing Intelligence: Origins, Obstacles, and Innovations* (Washington, DC: Georgetown University Press, 2008), p. 73.

21 Hilsman, Roger Jr. "Intelligence and Policymaking in Foreign Affairs". *World Politics* 5(1), (October 1952):34–35; Kent, Sherman. "Estimates and Influence". *Studies in Intelligence* 12(3), (1968):18–19; and Gregory F. Treverton, *Reshaping Intelligence for an Age of Information* (New York, NY: Cambridge University Press, 2003), p. 181.

22 Richard K. Betts, *Enemies of Intelligence: Knowledge and Power in American National Security* (New York, NY: Columbia University Press, 2007), p. 24.

23 James B. Steinberg, "The Policymaker's Perspective: Transparency and Partnership," in Roger Z. George and James B. Bruce, eds., *Analyzing Intelligence: Origins, Obstacles, and Innovations* (Washington, DC: Georgetown University Press, 2008), pp. 83–84.

DOI: 10.1057/9781137523792.0011

24 Thomas Fingar, *Reducing Uncertainty: Intelligence Analysis and National Security* (Stanford, CA: Stanford University Press, 2011), pp. 15–16, and 112.
25 Douglas J. MacEachin, "The Tradecraft of Analysis," in Roy Godson, Ernst R. May and Gary Schmitt, eds, *U.S. Intelligence at the Crossroads: Agendas for Reform* (Washington, DC: Brassey's, 1995), p. 75; and Michael Lewis, "Obama's Way," *Vanity Fair*, October 2012, http://www.vanityfair.com/politics/2012/10/michael-lewis-profile-barack-obama (Accessed on September 17, 2012).
26 Richard K. Betts, *Enemies of Intelligence: Knowledge & Power in American National Security* (New York, NY: Columbia University Press, 2007), pp. 159–182; Richard Clarke et al., *Liberty and Security in a Changing World* (December 12, 2013), available at: http://www.whitehouse.gov/sites/default/files/docs/2013-12-12_rg_final_report.pdf (accessed on: September 16, 2014); and Mundie, Craig. "Privacy Pragmatism". *Foreign Affairs* 93(2), (March/April 2014):28–38.
27 Wayne Michael Hall and Gary Citrenbaum, *Intelligence Analysis: How to Think in Complex Environments* (Santa Barbara, CA: Praeger Security International, 2010); and Long, Letitia A. "Activity Based Intelligence: Understanding the Unknown," *The Intelligencer* 20(2), (Fall/Winter 2013):7–15.
28 Karl Popper, *The Poverty of Historicism* (New York, NY: Routledge, 2002).
29 Richards J. Heuer, Jr., "Adapting Academic Methods and Models to Governmental Needs," in Richards J. Heuer, Jr., ed., Quantitative Approaches to Political Intelligence: The CIA Experience (Boulder, CO: Westview Press, 1978), pp. 4–5; and Gaddis, John Lewis "History, Theory, and Common Ground," *International Security* 22(1), (Summer 1997):75–85.
30 Richard J. Zeckhauser, "Investing in the Unknown and Unknowable," in Francis X. Diebold, Neil A. Doherty and Richard J. Herring, eds., *The Known, the Unknown, and the Unknowable in Financial Risk Management* (Princeton, NJ: Princeton University Press, 2010), pp. 304–343.
31 For many reasons, producers and consumers have been skeptical about the ability of mathematical models to capture the salient features of long-term, adaptive strategic competition. Early efforts to apply formal modeling in intelligence analysis were met with mixed results given the complexity of intelligence problems, the availability and quality of data and relevant theory, and a willingness of analysts to engage with methodologists in an effort to improve the rigor and transparency of assessments. Many of the concerns echo those that were raised during the formalization of Department of Defense planning and policy analysis during the 1960s and 1970s, which ultimately led to the creation of the Office of Net Assessment, and the development of alternative analytic approaches specifically tailored to cope with long-term uncertainties and adaptive behaviors in the examination of strategic balances and competition. Eliot A. Cohen,

DOI: 10.1057/9781137523792.0011

"Net Assessment: An American Approach," *Jaffe Center for Strategic Studies Memorandum No. 29* (Tel Aviv, IL: Tel Aviv University, 1990); Stephen Peter Rosen, "Net Assessment as an Analytical Concept," in Andrew W. Marshall, J. J. Martin, and Henry S. Rowen, eds., *On Not Confusing Ourselves: Essays on National Security Strategy in Honor of Albert and Roberta Wohlstetter* (Boulder, CO: Westview Press, 1991), pp. 283–330; and Bracken, Paul. "Net Assessment: A Practical Guide," *Parameters* 36, (Spring 2006): 90–100.

32 See *The Sims*, http://www.thesims.com/en-us; and *Spore*, http://www.spore.com/. Also see John E. Mayfield, *The Engine of Complexity: Evolution as Computation* (New York, NY: Columbia University Press, 2013).

33 Joshua M. Epstein, *Generative Social Science: Studies in Agent-Based Computational Modeling* (Princeton, NJ: Princeton University Press, 2006).

34 It is important to note the rudimentary characteristics of artificial laboratories given that all models are simplifications of systems and therefore do not contain the full-richness of the real-world they represent. Therefore, they are better considered tools for generating and supporting analytic inferences than crystal balls that replicate the real-world in all of its detail. See Joshua M. Epstein and Robert Axtell, *Growing Artificial Societies: Social Science from the Bottom Up* (Cambridge, MA: MIT Press 1996).

35 Greg Miller, "Oval Office iPad: President's Daily Intelligence Brief Goes High-Tech," *The Washington Post*, April 12, 2012, http://www.washingtonpost.com/blogs/checkpoint-washington/post/oval-office-ipad-presidents-daily-intelligence-brief-goes-high-tech/2012/04/12/gIQAVaLEDT_blog.html (Accessed on September 28, 2013); and Meador, Lawrence C. and Cerf, Vinton G. "Rethinking the President's Daily Brief". *Studies in Intelligence* 57(4) (December 2013): 1–14.

36 For examples see Meador, Lawrence C. and Cerf, Vinton G. "Rethinking the President's Daily Brief". *Studies in Intelligence* 57(4), (December 2013):1–14.

37 Intelligence scholars and practitioners have long debated the merits of new models of producer-consumer relations with some favoring increased closeness with consumers while others remain deeply skeptical of new models and argue that more should be done to return to established practices. For examples see Kerbel, Joshua and Olcott, Anthony. "Synthesizing with Clients, Not Analyzing for Customers". *Studies in Intelligence* 54(4), (December 2010):11–27; Paul R. Pillar, "The Perils of Politicization," in Loch K. Johnson, ed., *The Oxford Handbook of National Security Intelligence* (New York, NY: Oxford University Press, 2010), pp. 472–484; Jennifer E. Sims, "Decision Advantage and the Nature of Intelligence Analysis," in Loch K. Johnson, ed., *The Oxford Handbook of National Security Intelligence* (New York, NY: Oxford University Press, 2010), pp. 389–403; and Thomas Fingar, *Reducing Intelligence: Intelligence Analysis and National Security* (Stanford, CA: Stanford University Press, 2011).

DOI: 10.1057/9781137523792.0011

38 Paul Wolfowitz, "Comments: Paul Wolfowitz," in Roy Godson, Ernest R. May and Gary Schmitt, eds., *U.S. Intelligence at the Crossroads: Agendas for Reform* (Washington, DC: Brassey's, 1995), pp. 75–80.

39 Richards J. Heuer, Jr., "The Evolution of Structured Analytic Techniques," *Presentation to the National Academy of Science, National Research Council Committee on Behavioral and Social Science Research to Improve Intelligence Analysis for National Security* (Washington, DC December 8, 2009), available at: https://www.e-education.psu.edu/drupal6/files/sgam/DNI_Heuer_Text.pdf (accessed on September 15, 2014); Randolph H. Pherson and Richards J. Heuer, Jr., "Structured Analytic Techniques: A New Approach to Analysis," in Roger Z. George and James B. Bruce, eds., *Analyzing Intelligence: National Security Practitioners' Perspectives* (Washington, DC: Georgetown University Press, 2014), pp. 231–248; and Richards J. Heuer Jr. and Randolph H. Pherson, *Structured Analytic Techniques for Intelligence Analysis, 2nd Edition* (Washington DC: CQ Press/SAGE Publications, 2015).

40 For a discussion of models as both collaborative and analytic tools see Edward Waltz, *Quantitative Intelligence Analysis: Applied Analytic Models, Simulations and Games* (Lanham MD: Rowman-Littlefield, 2014), pp. 21–23.

DOI: 10.1057/9781137523792.0011

Index

The 'f' and 'n' after the page numbers indicate figures and notes, respectively.

DOI: 10.1057/9781137523792.0012

DOI: 10.1057/9781137523792.0012

DOI: 10.1057/9781137523792.0012

DOI: 10.1057/9781137523792.0012

DOI: 10.1057/9781137523792.0012

CPSIA information can be obtained
at www.ICGtesting.com
Printed in the USA
LVHW092201200120
644238LV00007B/287

9 781137 523785